THE
CANAL
GUIDE

BRITAIN'S 50 BEST CANALS

STUART FISHER

ADLARD COLES NAUTICAL

BLOOMSBURY
LONDON · NEW DELHI · NEW YORK · SYDNEY

Published by Adlard Coles Nautical
an imprint of Bloomsbury Publishing Plc
50 Bedford Square, London WC1B 3DP

www.adlardcoles.com

Bloomsbury is a trademark of Bloomsbury Publishing Plc

ISBN 978-1-4729-1852-9
ePDF 978-1-4729-1854-3
ePub 978-1-4729-1853-6

A CIP catalogue record for this book is available from the British Library.
This book is produced using paper that is made from wood grown in managed, sustainable forests. It is natural, renewable and recyclable. The logging and manufacturing processes conform to the environmental regulations of the country of origin.

Contains material previously published in *Canals of Britain 2nd Edition* by Stuart Fisher, Adlard Coles Nautical, 2012.

Typeset in Gothic 720
Book design by CE Marketing
Printed and bound in China by C&C Offset Printing Co

Note: while all reasonable care has been taken in the publication of this book, the publisher takes no responsibility for the use of the methods or products described in the book.

10 9 8 7 6 5 4 3 2 1

Legend for maps

Featured canal
Other canal or river
Motorway
Other road
Railway
Open water or sea
Inter-tidal zone
Built-up area
Woodland

Contents

Introduction

We are fortunate to have, in this country, a canal network like no other in the world. It was the first commercial canal system, leading the way for the Industrial Revolution, but has remained largely as it was originally built. The canals are mostly small and intimate. Restoration in recent years, supported by lottery and other funding, has outstripped the pace of construction even during the 'canal mania' years.

Overseas, where canals have been enlarged to take modern commercial craft, you can look at the distant bank and wonder whether you will be run down by something the size of an office block. In this country the biggest risk is running aground and you can usually wade through the mud to either bank; a canal is a safe environment in which there is limited scope for getting into serious trouble. For the walker and cyclist, canals provide routes that are mostly flat. As far as possible, descriptions are given downhill and with the flow for those who have a choice of direction.

We have canals with scenery that changes frequently: open countryside, wildlife, heritage industrial buildings, canalside public houses, modern city centres, wild moorland and coastal harbours, all mixed up. Anywhere on the system fantastic engineering structures can be found.

For those with suitable boats there are 65,000km of navigable river in England and Wales alone; scope for a whole library of books, but in this one I have concentrated on Britain's 50 best canals. Not all of them are linked to the rest of the system and not all are physically passable for many boats or have useable towpaths. You may need a spirit of adventure, like the earlier recreational boaters.

Who uses the canals? If you look at canal magazines you will see smiling couples or families busy in the summer sunshine. In practice, you may find the picture rather different. Often the canals are deserted, except for wildlife that finds them an ideal environment, without needing all humans and boats to be banned.

The intention is that this book should be engaging to all who travel the canals. I do not usually give navigation instructions, depths and headrooms, portage routes or what to do when the towpath runs out. If the present state of a canal is such that it is limited to one kind of user, usually someone able to undertake portages, I may refer to that category of user, otherwise I talk more generally. I draw attention to features near the canal, especially in heritage cities such as Bath and Chester, because most canal travellers will not want to pass through these without stopping.

Much in the future will depend on volunteers. As walkers and cyclists are often the major users of canal routes the Canal & River Trust (CRT) will need to have a more positive approach towards them, including signposting routes at

tunnels and other places where towpaths do not follow the canal, removing barriers from towpaths and ensuring routes connect rather than making just token sections available.

Along the canals there are some consistent trends. Heavy industry is evaporating. Public houses across the country are closing at a rapid rate, canalside pubs included, although a number have been converted to restaurants. Virtually anything which can be converted to housing, including warehouses and other former industrial premises, is now packed with residential occupants.

The canals may be ever-changing but there's a sign that more and more people are rediscovering them and taking them to their hearts – in recent years canal holiday bookings have shot up.

Birmingham Canal Navigations: Old Main Loop Line

Possibly because Birmingham is the only major city not located on a large river, it has had to rely on its man-made waterways, having more canals than Venice. The whole canal network spreads out from Birmingham and it is to the Birmingham Canal Navigations (BCN) that all the loose ends connect. It is, therefore, intensely complex, completely built-up and industrial. Its commercial influence is declining but remains sufficient to reduce its attraction as a cruising waterway, resulting in lighter traffic.

Of Britain's operational canals, only the Fossdyke and the Bridgewater pre-date the Birmingham. The Birmingham Canal Company were authorised to build their line in 1768 and the following year Brindley opened it as far as Wednesbury. The whole 36km course was completed in 1772 and, because of the minerals and industry along its route, it was immediately highly successful.

Many of the loops of the old line have since been cut off or filled in, but the section between Tipton and Smethwick remains, and is a contour canal with more features of interest than are found along the more efficient New Main Line. The old line divides from the new at Tipton Factory Junction, just a stone's throw from the top lock on the New Main Line.

Tipton Green has been pleasantly landscaped to make the most of the canal, with assorted new housing standing around the Fountain Inn. This was formerly the home of 19th century canalman and prize fighter William Perry, known as the Tipton Slasher, and his statue stands on the green. The Malthouse Stables at

The Malthouse Stables centre at Tipton Factory Junction. Only the two-storey part is original.

Tipton have been restored as an outdoor activity centre. From Tipton Junction there is a spur – the Dudley Canal No 1 Line – which leads to the Black Country Museum and the Dudley Tunnel.

The houses, with their gardens hiding behind greater reedmace, give way to old brick factories. The view opens out at Burnt Tree with new housing around a marina, and the Silurian limestone ridge behind it. The canal is wider here as it approaches Dudley Port. A container yard on the left is an update of the area's former purpose.

The canal crosses the Netherton Tunnel Branch Canal on an aqueduct and it is worth looking over the parapet at the imposing cutting leading to the mouth of this 3km long tunnel.

At Brades Hall Junction there are two arches on the left, the first abandoned but the second leading past an ivy-covered wall to the first of the Brades Locks on the Gower Branch Canal. This drops down as a midway connection to the New Main Line. Once again it is an area of industry, with factories alongside the canal.

The character of the canal changes abruptly at Oldbury as the M5 is in close proximity to

Distance

10km from Tipton Factory Junction to Smethwick Junction

Highlights

Telford's listed Steward Aqueduct, where motorway passes over railway, which passes over canal, which passes over canal

Oldest working locks in Britain, at Spon Lane

Galton Valley Canal Heritage Area

Telford's Engine Arm Aqueduct

Navigation Authority

Canal & River Trust

Canal Society

Birmingham Canal Navigations Society

www.bcnsociety.co.uk

OS 1:50,000 Map

139 Birmingham & Wolverhampton

the canal for the next 2km, mostly overhead. As with the River Tame in Birmingham, an elevated route over a waterway has proved to be the most acceptable line for a motorway to be squeezed through the city.

North Junction and South Junction of the closed Oldbury Loop Line are still visible and the Houghton Branch Canal leads away under the motorway. Once under what can be a useful canopy for a rainy day, Oldbury Locks Junction accepts the Titford Canal, which descends through six locks from the Crow, a feeder from Rotton Park Reservoir.

The increasingly complex concrete jungle intensifies as the A457 passes between canal and motorway, and the canal comes out into the open air for a breather. At one point there is a traditional brick-arched bridge over the canal, out of place among all the vast concrete columns and walls. The complexity reaches its zenith as the M5 passes over the main railway

The Gower Branch Canal drops away from Brades Hall Junction through the Brades Locks.

Smethwick Lock, formerly duplicated. The chamber on the left side has been filled in.

line, which passes over the Old Main Loop Line, which in turn passes over the New Main Line on the Steward Aqueduct. (The aqueduct would probably have been more of an honour to BCNS committee member Stewart if the name had been spelled correctly.) Telford's 2.1m iron-trough aqueduct of 1826–1828 is now a listed structure.

Spon Lane Wharf and Junction are now beneath the motorway. Spon Lane Locks Branch provides a connection down to the New Main Line through Spon Lane Locks, the remaining bottom three from the six that descended from the original summit and probably the oldest working locks in Britain. Top Lock has a split cantilevered bridge through which ropes could be passed without towing horses being unhitched.

The site of one of Britain's greatest concentrations of canal architecture, the cutting between Sandwell and Smethwick has been designated the Galton Valley Canal Heritage Area. Chance's former glassworks on the right were founded in 1824 and include a number of listed buildings. Chance pioneered sheet

glass, produced optical glass for lighthouses after 1838 and made the glass for the Crystal Palace in 1851, before going on to manufacture microscope lenses, rangefinders, telescopes and searchlights.

Looking back on the left side the prominent feature is the listed small timber belltower of 1847 on the seven-storey offices of Archibald Kenrick & Sons, who have made ironmongery since 1791. George Salter, manufacturers of such things as spring balances, weighing equipment and steam locomotive safety valves, sited their foundry beside Top Lock.

Eventually the M5 turns away and the canal enters a deep cutting below the original summit. Until they were dismantled in 2006, colliery loading chutes were notable on the left bank. Built in around 1930, they were fed by a narrow tramway from the Sandwell Park and Jubilee collieries, but were later replaced by conveyors. While boats were being loaded, boatmen and horses sheltered in a brick building, the ruins of which still stand on the opposite bank.

Built in 1791 and a scheduled ancient monument, Summit Bridge is a great brick arch with an unusual sloping parapet wall. The concentric arches reduce towards Galton Tunnel, which has a towpath and passes under the A4168 Telford Way. Samuel Galton was a BCNS committee member, self-educated in the sciences and the owner of a gun foundry.

Smethwick New Pumping Station was built in 1892 and had two steam engines to pump water from the New Main Line to the Old Main Loop Line in order to replace water lost by boats locking down at Spon Lane and Smethwick. Intended to supersede the Smethwick Engine, it ceased operation in the 1920s although a diesel engine was installed for fire fighting during the Second World War. It is a restored listed building, now used as a museum.

The Engine Arm was a feeder to the summit level from Rotton Park Reservoir via the Boulton & Watt engine, which operated for 120 years. It leads across Telford's magnificent Engine Arm Aqueduct of 1825, a scheduled ancient monument, past the Galton Valley Canal Heritage Centre. The Old Main Loop Line towpath travels over the Engine Arm Canal on a brick footbridge with indented honeycomb stone quoins.

The three locks down to the lower level are the bottom ones of the original six, all listed buildings. In 1789 Smeaton duplicated these three locks. Brindley's originals were filled in during the 1960s. There was a Toll House between the upper pair. The layout of Pope's Bridge, carrying Bridge Street, shows the alignment of the two lock flights.

The two main lines meet at Smethwick Junction, Soho, notable features being two cast-iron footbridges installed in 1828 after being prefabricated at the Horseley Ironworks in Tipton. Once again these are listed buildings. Their semi-elliptical shape gives an advantage over segmental curves by allowing greater headroom for horses passing below.

Birmingham Canal Navigations: New Main Line

The major improvement to the Old Main Line came between 1825 and 1838 when Telford engineered the New Main Line between Deepfields and Birmingham. He introduced bold cuttings and embankments, producing extra water space to ease traffic congestion and shorten the route by 11km at the lower 138m Birmingham Level.

The canal starts from Gas Street Basin at the end of the Worcester & Birmingham Canal in the centre of Birmingham. Dating back to 1821, the Tap & Spile pub precedes Broad Street Bridge, a vast tunnel with buildings

The Broad Street Bridge.

The International Convention Centre at Brindleyplace.

Distance
24km from Gas Street Basin to Aldersley Juntion
Highlights
Brindleyplace, with its hub of bars and
restaurants, as well as Symphony Hall and the
International Convention Centre
Old Turn Junction, with the National Indoor
Arena and National Sea Life Centre
The imposing cast-iron Telford Aqueduct
The Netherton Tunnel – the largest cross-section
canal tunnel in the country
The 21-lock Wolverhampton Flight
Navigation Authority
Canal & River Trust
Canal Society
Birmingham Canal Navigations Society
www.bcnsociety.co.uk
OS 1:50,000 Map
127 Stafford & Telford
139 Birmingham & Wolverhampton

on top of it. Much of the New Main Line has twin footpaths, partly because of the complexity of the system and partly to reduce congestion.

The Pitcher & Piano and the Handmade Burger Company are among the amenities at the much-restored Brindleyplace opposite the International Convention Centre and Symphony Hall. Much of the blue brickwork on the towpath bridges has been renovated and the towpaths converted into attractive canalside walks, now fully surfaced with brickwork.

Old Turn Junction is at the start of the Birmingham & Fazeley Canal, opposite Sherborne Wharf on the Oozells Street Loop. Three of the four corners of the junction are occupied by the Malt House hostelry (famously visited by Bill Clinton), the National Sea Life Centre and the National Indoor Arena.

Three loops show where the contour canal used to run before being straightened: Oozells Street Loop, Icknield Port Loop via Rotton Park and Soho Loop, which winds its way past the prison at Winson Green. The Soho Loop rejoins at the first of several toll islands on the canal. By this stage it has already been joined by the West Coast Main Line, remaining in close proximity for most of the distance. Another arrival, and a feature of this part of the canal system, is the presence of purple lupins growing wild on the

embankments. Bridges over side arms have low lattice parapets, manufactured in the local Horseley works.

At Smethwick the Old Main Loop Line diverges to the right, rising through three locks from the Birmingham Level to the Wolverhampton Level. A feeder, Engine Branch (named after the Boulton & Watt steam pumping engine that fed the Birmingham Canal Navigations summit level for 120 years), crosses the New Main Line on Telford Aqueduct.

Telford Aqueduct carries the Engine Arm over.

The aqueduct is a magnificent cast-iron structure, highly decorated, its dark brown paintwork highlighted with red and white detailing.

The New Main Line turns into its boldest cutting at Sandwell. When they were undertaken, the earthworks here were some of the greatest in the world. On the north bank is Galton Valley Canal Heritage Centre. There was formerly a clear view straight down the 21m deep cutting to Telford's elegant Galton Bridge. A fitting end to the straight cut, it is 23m high and has a 46m span – the world's longest canal span when it was built in 1829. In 1974 this changed with the construction of the 112m long Galton Tunnel (which has a towpath) next to the bridge, with an embankment over the top to carry the A4168. With a railway bridge just beyond, Galton Bridge can no longer be seen at its best from either direction.

A long straight section of the canal passes under another interesting group of bridges. The Old Main Loop Line crosses to the higher ground on the south on the Steward Aqueduct, and the

M5 viaduct is supported on uncompromising nodes in the centre of the canal.

Pudding Green Junction leads off northwards to the Walsall Canal in an area of small but bustling works. By now the New Main Line is on a dead-straight 4km run through to Tipton. At Albion Junction the Gower Branch, with its deep locks, connects with the Old Main Loop Line. The only extensive views on this section are of the hills around Dudley.

From Dudley Port Junction the Netherton Tunnel Branch runs parallel to the Gower Branch and, in the distance, it can be seen passing under the Old Main Loop Line and up to the mouth of Netherton Tunnel, the largest cross-section canal tunnel in the country.

Residential properties close in on the south side of the canal and a canal cottage sits on top of the embankment near a couple of aqueducts. The Ryland Aqueduct of 1968 clears the A461 with a single 24m concrete span.

Beyond the Noah's Ark pub, the three Factory Locks bring the New Main Line up to rejoin the Old Main Loop Line at Tipton Factory Junction. Noteworthy are a split bridge over the bottom lock, a boatman's chapel now converted into a factory, a large warehouse and a BCN cast-iron boundary post. The Factory Bridge of 1825 has now been relocated to the Black Country Living Museum in Dudley. The Staffordshire Thick Coal seam was 15m thick here and near the surface.

At Deepfields Junction the Wednesbury Oak Loop is a remainder of the contour canal. Contours changed dramatically at Deepfields on one occasion when subsidence dropped the canal area by 1.2m within a few hours. Factories lie in various stages of dereliction. Bilston Steelworks was, until 1981, the last surviving blast furnace in the Black Country. Yellow iris is often present along the banks.

As the canal edges into Wolverhampton, capital of the Black Country, factories appear in

Galton Bridge, now less easy to view.

The approach to Wolverhampton top lock.

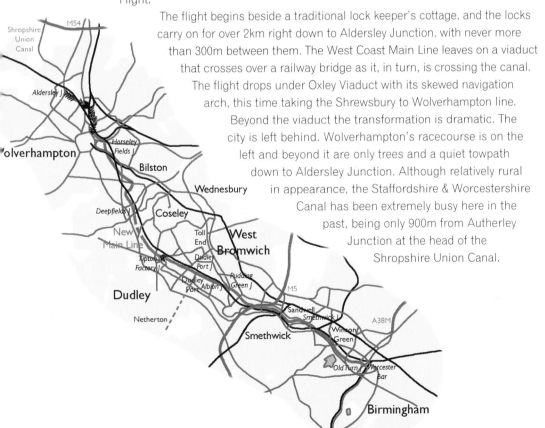

quick succession. A large tunnel with twin towpaths supports a multistorey carpark, followed by a Canal & River Trust depot with barge entry doors in the side. The basin beyond it is the jewel of the canal – an attractively laid out garden area with benches and narrowboats moored at the top of the 21-lock Wolverhampton Flight.

The flight begins beside a traditional lock keeper's cottage, and the locks carry on for over 2km right down to Aldersley Junction, with never more than 300m between them. The West Coast Main Line leaves on a viaduct that crosses over a railway bridge as it, in turn, is crossing the canal. The flight drops under Oxley Viaduct with its skewed navigation arch, this time taking the Shrewsbury to Wolverhampton line. Beyond the viaduct the transformation is dramatic. The city is left behind. Wolverhampton's racecourse is on the left and beyond it are only trees and a quiet towpath down to Aldersley Junction. Although relatively rural in appearance, the Staffordshire & Worcestershire Canal has been extremely busy here in the past, being only 900m from Autherley Junction at the head of the Shropshire Union Canal.

3 Wyrley & Essington Canal

The Wyrley & Essington Canal connected the Birmingham Canal Navigations' Main Line at Wolverhampton with the Coventry Canal at Huddlesford, forming the BCN's most northerly loop and today displaying its most attractive scenery, almost all on one level. The canal follows a tortuous line around the contours, earning it the nickname Curly Wyrley. The locks, which served various coalfields, are all positioned on the numerous branches.

The line leaves the BCN Main Line at Horseley Fields Junction. Light traffic on it means that it is remarkably clear, except for the waterweeds. The West Coast Main Line crosses immediately and the canal then goes back a couple of centuries as it passes a constriction caused by what was once a toll island for coal barges in the centre of the canal. Tunnels under the towpath carry former spurs that served adjacent factories.

After the blackened spire at Heath Town and the Jolly Collier pub, the canal passes playing fields and grassed areas, and a modern church with a striking green roof. The old brick bridge leading up to the hospital has been repaired very obviously and carries a nameplate, as do the other bridges on this canal.

Beyond it, a substantial lattice bridge, of a pattern we will see again later, carries the towpath at Wednesfield Junction (which used to have an island with an octagonal tollbooth) across the Bentley Canal. This was abandoned as a through route to Bentley in 1961, and the buildings of Willenhall have spread over it in recent years.

A path through the lilies at Harden.

Remains of the toll island.

A red sandstone towered church lends dignity to an area where the Royal Tiger and Spread Eagle pubs flank the canal, together with a school. Houses have their gardens backing on to the left bank, often making special features out of their canalside locations.

At Short Heath, the only wooded area on the whole canal is enfolded in a large sweep as the route turns northwards. Gradually, the roar of traffic becomes audible as the M6, one of the country's busiest stretches of road at this point, pulls alongside and crosses over.

The Sneyd & Wyrley Bank Branch climbs up from Sneyd Junction towards Great Wyrley but does not get far before becoming derelict. Meanwhile, the main line doubles back on itself to cut down the other side of the valley.

Birchills Junction takes the Walsall Branch Canal away as the main line turns north once

Distance
34km from Horseley Fields Junction to Huddlesford Junction

Highlights
Beautiful scenery from Little Bloxwich to Pelsall
The Swan – a pub in Pelshall, haunted by a coal-covered miner

Navigation Authority
Canal & River Trust

Canal Societies
Birmingham Canal Navigations Society
www.bcnsociety.co.uk
Lichfield & Hatherton Canals Restoration Trust
www.lhcrt.org.uk

OS 1:50,000 Map
(127 Stafford & Telford)
128 Derby & Burton upon Trent
139 Birmingham & Wolverhampton

again. Ill feeling and rivalry between companies prevented the connection from being made until 1840 when the Wyrley & Essington Canal and BCN were amalgamated and came under railway control. Unusually, the railway encouraged traffic on this canal, building interchange basins. Consequently, the Wyrley & Essington saw some of the last commercial traffic on the BCN. The end came suddenly in 1966, when coal transport ceased.

Moving away past playing fields and a church with a strange little square green spire, the canal passes clumps of watercress, which indicate that the water is fairly clean.

At Harden a higher-level section slips past tree-lined hospital grounds and clears the housing of Little Bloxwich by the Barley Mow pub. It then breaks out into open country, the most attractive stretch of the whole canal, between Little Bloxwich and Pelsall. Having come down from Wolverhampton, it is immensely satisfying to pause by the farm at Fishley, the late-afternoon sun shining over one shoulder, and listen to just the distant rumble of traffic and the closer cawing of rooks, the only sounds to break the silence.

Bridges carry the towpath over factory feeders near Horsley Fields Junction.

The Lord Hay's Branch to Newtown has now been lost but over 2km of the Cannock Extension can be seen running away from Pelsall Junction in a dead-straight line.

After passing the Royal Oak pub and some new houses, the route becomes especially tortuous, with the bends being replaced with corners that are tight for narrowboats. The Swan pub (along with its ghost of a coal-covered miner

Cast-iron bridge over the canal at Ogley Junction, the current foot of the canal.

in the bar) stands away behind a house with a prominent barrel-shaped dovecote.

Catshill Junction brings in the Hay Head or Daw End Branch, the most important branch of all as it linked up with the Rushall Canal to provide access to the industry of Birmingham. As the Wyrley & Essington loops round Brownhills, it passes the Anchor pub and then reaches open country on the right, with extensive views in the direction of Shenstone.

At Ogley Junction the Anglesey Branch joins, bringing water supplies down from Chasewater. A lattice bridge takes the towpath across and the Wyrley & Essington turns right into a basin that is now the effective terminus of the canal. The canal was extended to Huddlesford in 1797 via the 30 locks of the Ogley Flight but this was abandoned in 1954 and now a garden centre stands on what was once a significant engineering feature.

The canal has lost 10km from here. The route formerly ran to the north of the prominent church spire at Wall and past the more famous spires of Lichfield. There are plans to restore this section as the Lichfield & Hatherton Canal.

The Cannock Extension leaves Pelsall Junction.

Now just 500m remains at the far end, the south-west end of which is a marina with entry barred (although a public bridleway crosses the stump). The Wyrley & Essington Canal then joins the Coventry Canal by the Plough pub at Huddlesford, a small community cut in two by the railway.

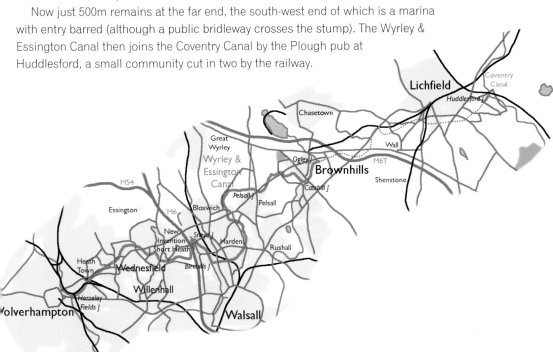

Walsall Canal

The Walsall Canal links the Wednesbury Old Canal at West Bromwich with Walsall in the West Midlands. Construction began in 1786, although the centre of Walsall was not reached until 1799, after this relatively short and straightforward canal had curved its way round Wednesbury.

The canal has eight locks, all of which come immediately at the Ryders Green Flight. At the southern end the waterway leaves the Wednesbury Old Canal and passes the Eight Locks public house by the top lock.

This canal is more free of weeds than are some others. The marshy area to the right by the bottom lock is predominantly greater reedmace but toadflax gets a grip, with its pale yellow flowers, where dry ground is available. Opposite is the point where the Haines Branch entered.

The Walsall Branch Canal leaves via Walsall Locks beyond Walsall Junction.

An old railway bridge crosses at Toll End and, soon after, the Lower Ocker Hill Branch – or the remaining 300m of it – leaves on the left next to tennis courts and other sports facilities.

The Tame Valley Canal leaves on the right between a couple of the lattice cast-iron bridges that are among the attractive features of the Black Country canals.

A canal cross exists at the point where the former Gospel Oak Branch joins. Presumably this must have given some logistical problems with the movement of horses and towlines at busy times.

The first line of the Midland Metro crosses. The Monway Branch has disappeared under the earth but the banks of the Walsall Canal itself remain untouched and lupins and orchids may be found. The water is ochre-coloured but remains relatively clear because it is only lightly used and,

Distance
11km from Wednesbury Old Canal to Walsall
Highlights
Ryders Green Flight
Walsall Public Wharf and the New Art Gallery
Jerome K Jerome Birthplace Museum
Navigation Authority
Canal & River Trust
Canal Society
Birmingham Canal Navigations Society
www.bcn-society.co.uk
OS 1:50,000 Sheet
139 Birmingham & Wolverhampton

like sections with other less-than-inviting colours of water, minnows can be seen swimming about in significant numbers.

Powerlines now follow the canal all the way to Darlaston Green. From Bilston to Darlaston Green the Black Country Route flanks the canal.

The Bradley Branch Canal, now heavily overgrown with reeds, leaves to the left at Moorcroft Junction. The reeds provide a haven for moorhens.

The Moxley Stop was near the red sandstone spire of the church in Moxley. Today, a children's playground is more prominent.

As the canal moves on past the lines of the former Bilston Branch and Willenhall Branch, the countryside comprises derelict fields, occasionally occupied by horses. A notable feature of the Black Country canals is that the people always seem friendly, whereas the common experience elsewhere is that country people are more sociable. Black Country friendliness will be encountered on this canal and must be counted as one of its assets.

After a school and then Bug Hole Wharf in the vicinity of an electricity substation and more schools, the canal runs through a section with

Unusual cemetery entrance off the towpath.

factories but also with newer canalside houses, which, from their expensive leaded windows, appear to be aimed at the more affluent end of the housing market.

The powerlines leave at Darlaston Green Wharf and the canal curves round to cross an aqueduct over the railway, immediately before the weed-choked remains of the Anson Branch. Over the next reach the canal is on a low embankment with a cemetery to the south surrounding a church with a small spire. A little further on are two very large gas holders next to the M6, but it is the busy motorway itself that is now prominent. Aqueducts take the canal over a minor road and the fledgling River Tame, before the canal passes under the M6.

In this vicinity, the whole area is riddled with old coal mines. Heavy metal pollution from former copper refining on the derelict site to the left of the canal has been a major problem in the past, as drainage from polluted ground went into nearby headwaters of the River Tame. The former factory here was one of two that were said to account for 18 per cent of the copper and 17 per cent of the nickel found in the water at the Tame/Trent confluence.

The canal then moves on into the area of Pleck where a Sikh temple stands grandly next to the water. After a public house by a bridge the route passes a former canal wharf building with the hoist points and doors at various levels, still obvious above the water. High fences and

The restored canal building at Walsall.

Sikh temple overlooking the canal at Pleck.

industrial premises front a cemetery with an interesting circular entrance gateway.

After some recent housing by the canal, the Walsall Branch Canal leaves up Walsall Locks from Walsall Junction. This important link with the Wyrley & Essington Canal was not made until 1841 because of company rivalry but today it cuts out a significant detour via Wolverhampton for boats going north.

Walsall Public Wharf is now a grassed area where a bar has waterside seating facing a canal terminus building, which has been restored. High blocks, including the New Art Gallery, and glass walls rise behind. Walsall has changed much since it took its name from the Old English for Walh's valley, 'Walh' meaning a Welshman. In the 1890s it saw the activities of the Walsall anarchists.

A different aspect of boating is featured in the Jerome K Jerome Birthplace Museum.

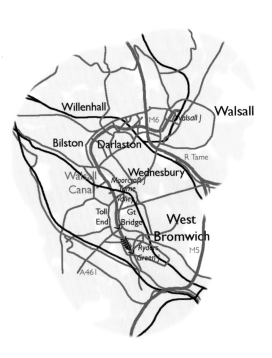

5 Daw End Branch

Also known as the Hay Head Branch, the Daw End Branch of the Wyrley & Essington Canal was cut in 1800 as a narrow contour canal to take out limestone. The wandering route around Aldridge allowed it to be lock-free.

The branch leaves the Wyrley & Essington main line at Catshill Junction on the edge of Brownhills. Narrows were a toll control point.

Clayhanger Bridge is a modern structure near the former Clayhanger Wharf. Pools at Clayhanger have marsh orchids while the canal has yellow irises and yellow lilies, among which coots dabble. The Beacon Regional Park follows through what had been part of Cannock Forest.

Black Cock Bridge crosses by the site of Black Cock Wharf, a single-lane bridge with uniform slopes up to a sharp summit, a bit like driving over the roof

Dog roses, yellow iris and white-and-yellow lilies on the Rushall Canal.

of a house. It is a blind approach for drivers so there is constant warning hooting.

The banks have been built up because coal-mining subsidence has affected the surrounding area. Indeed, a pair of 1948 semi-detached houses were braced together after they settled.

Plenty of anglers can be found here and there is a large rust-coloured silhouette of one next to Walsall Wood Bridge, which takes the A461 over. Also adjacent is the site of the Eutopia brickworks, bricks from which were used for building up embankments.

Industrial premises with security fencing frequently stand by the canal but reedmace and broom soften the views and clumps of trefoil brighten up the stonework on the towpath side. Bream, carp, perch, roach and crayfish are in the water, while herons, swans and kingfishers also frequent it and flocks of noisy gulls wheel nearby. Minor bridges include the low Hollanders Bridge.

A sharp turn takes the canal westwards, away from industrial Aldridge, which was occupied in the Stone Age, as shown by flint finds. Although

Distance
13km from Wyrley & Essington Canal to Tame Valley Canal
Highlights
Beacon Regional Park
Hayhead Nature Trail
Navigation Authority
Canal & River Trust
OS 1:50,000 Sheet
139 Birmingham & Wolverhampton

the church of St Mary the Virgin is 19th century, it has a 14th century tower. The commercial sites include Anchor Brook Industrial Park. Aldridge was chosen as the site of Duckham's refinery and also had a cement works. Cement made here was retrieved from a sunken barge to repair sunken canal banks.

The original canal head line leads from Longwood Junction.

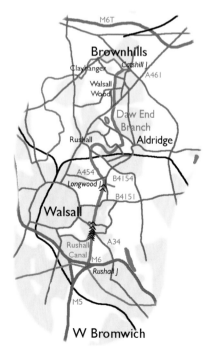

Catshill Junction where the branch leaves the main line.

At Rushall the canal turns south again. The Boathouse pub with mock lock gates faces the canal next to the B4154 crossing at Daw End. Adjacent is the Manor Arms, which dates from 1104 and has sold beer since 1248, over the years having been a mill, accommodation for monks and a farmhouse, but has no bar. During VE Day celebrations the chimney was struck by lightning, which was mistaken for a bomb. Glenn Miller was a regular customer and its Geriatric Club members have been mostly millionaires.

Park Lime Pits are designated as a nature reserve. Beyond the Railway Aqueduct, crossing the Walsall to Water Orton line, Rushall Hall lies to the west. It was attacked in 1643 by Prince Rupert and made a Royalist headquarters for a short period of time. The canal is now out into rural farmland.

Beyond Longwood Bridge, crossed by the A454 and Beacon Way, the head of the canal lies to the left, now moorings for Longwood Boat Club. It served Hayhead limeworks, which mined high-grade Silurian limestone and now has Hayhead Nature Trail.

Longwood Junction was set 600m from the head of the canal in 1847 to link the Rushall Canal, an almost straight cut to the south, running through the suburbs of Walsall, although the canal provides a green lane with little indication of the residential development that has taken place. This route offered a faster line to carry Cannock coal away to the south-east. White lilies and dog roses line the canal. There are clumps of elder bushes and the setting is occupied by Canada geese, moorhens and the occasional tern.

A moat and a golf course are located to the west of the first pair of locks, with only the B4151 crossing to disturb the following 2km straight pound. The remaining seven locks are spread over 900m, crossed by the A34 and with another golf course to the west.

The canal now moves from Walsall to Sandwell and traffic noise gradually builds up. The Beacon Way and sets of powerlines from a substation cross before the canal passes into the triangular junction 8 of the M6 where the M5 joins. The canal had its own junction here first, the Rushall Canal joining the Tame Valley Canal at Rushall Junction downwind of a large sewage works.

The Rushall Flight on a peaceful summer evening with the M6 traffic barely audible.

6 Tame Valley Canal

The Tame Valley Canal was one of the last canals to be built, running eastwards across the West Midlands from Tame Valley Junction, on the Walsall Canal, to Salford Junction, on the Birmingham & Fazeley Canal, which is at the head of the Grand Union Canal. Although planned earlier, it was not built until 1844, by which time the Old Turn to Salford Junction section of the Birmingham & Fazeley Canal was hopelessly congested with 24-hours-a-day, 7-days-a-week working. The Tame Valley Canal was constructed to bypass this section and was very sophisticated, with long, bold, straight sections that made use of high embankments and deep, wide cuttings, twin towpaths and brick-lined banks.

Other than tennis courts and other sports facilities beyond the Lower Ocker Hill Branch, the Tame Valley Junction area is inauspicious. Lattice-arched footbridges are an attractive feature of the Black Country canals.

The first reach of the canal is crossed by the Black Country Spine Road. Just before Golds Hill Wharf the canal is crossed by a former railway bridge with a pier on a central island in the canal.

The Midland Metro crosses at Golds Green.

There are plenty of flowers to brighten things up. Birdsfoot trefoil accompanies blackberries and there are no less than three shades of clover adding a splash of colour in summer.

A bridge arch carries the towpath over the entrance to Holloway Bank Wharf. This comes before a 3km straight that is closely followed by powerlines. Houses on the outskirts of West Bromwich are nearby on the right beyond long grass. More wild vegetation on the left stands at the top of a high embankment with a stream, pond, college and school at the foot

The land gradually rises again to Church Hill at Wednesbury. A short aqueduct carries the canal over a minor road before the canal runs into a deep cutting. This ends just before the large A4031 bridge. The canal emerges to cross an aqueduct over a minor road, with a public house below on the left, and then crosses another aqueduct over the railway.

An island divides the canal as it prepares to turn on to its most dramatic section, the kilometre-long embankment that runs high across the Tame Valley. The canal first crosses the western arm of the M5 on a concrete aqueduct. The height of the embankment can be judged by looking at the elevated section of the M6. This runs beside the canal but can be seen moving away to the north-west on spindly columns. A stop gate provides protection as the canal approaches the aqueduct that is to carry it across the River Tame.

At the far side of the valley the motorway climbs away up to Great Barr. The canal divides around another island as it approaches Rushall Junction, where it was joined by the Rushall Canal to win the coal trade from the Cannock mines. This is now a popular cruising link. The water quality in the Tame Valley Canal improves from here with the only visible wild plants being clumps of watercress along the edges.

The canal then passes under the eastern arm of the M5 and subsequently leaves the noise

Distance
14km from Tame Valley Junction to Salford Junction
Highlights
Midland Links triangle
Spaghetti Junction
Navigation Authority
Canal & River Trust
Canal Society
Birmingham Canal Navigations Society
www.bcnsociety.co.uk
OS 1:50,000 Sheet
139 Birmingham & Wolverhampton

and traffic behind. It dives into a deep, peaceful wooded cutting, where the pleasantly rural aspect belies the fact that the canal is passing between the housing estates of Grove Vale. Two bridges pass high across the cutting. The first carries the dual carriageway A4041 and the other is a slim footbridge carried on the piers of what looks like a former railway viaduct.

The kilometre-long cutting emerges on to a longer embankment, with extensive views over the Tame valley and back towards West Bromwich. After another island, one side blocked off by Hamstead Wharf, there are two aqueducts.

Roses grow on the lock cottage at Witton.

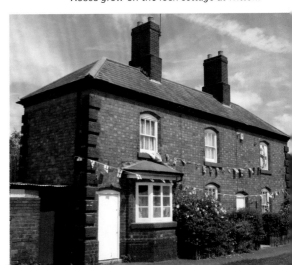

An island at Rushall Junction to control traffic using the Rushall Canal.

Passing TS *Leopard*, with its anti-aircraft gun, the canal plunges into another deep cutting. There are frequent outcrops of red sandstone. These are sufficiently soft for them to have needed to be reinforced with sections of red brickwork in many places. Meanwhile, the canal is edged with black bricks laid with black mortar, which provides a very smart finish. This is said to be one of the best angling waters in the Midlands.

The smartness is also present in the Perry Barr First Flight of locks, which were refurbished with attractive brickwork. Barr Top Lock Wharf lies beyond the A34 bridge. The first seven locks come as a 700m flight, on the way passing the Birmingham Alexander Stadium in Perry Park. It has an artificial football pitch, running track and full athletics facilities in a large, modern site.

The M6 crosses on the Thornbridge Viaduct, a viaduct that was extensively repaired using a very powerful water lance to cut out deteriorated concrete. Four more locks follow, forming Perry Barr Second Flight. Lock 11 is set between a container terminal and a sports ground lined with poplar trees, which shed their downy white seed on to the canal profusely in the autumn. The vicinity of Perry Barr Wharf has several industrial areas. Among these are two significantly older churches at Upper Witton, the first near the canal, with a rather short spire, and the other set amidst a cemetery.

The canal moves into a heavily industrial area for the final two locks, Perry Barr Third Flight. Beside the second of these is an old lock cottage covered with roses, defiantly holding on in its industrial surroundings. All is not derelict, however. As well as the lock cottage, there are various other places where the towpath can offer a colourful assortment of vetch, convolvulus and purple lupins, the latter thriving in the vicinity of Salford Junction.

The junction lies at the far end of Junction 6 of the M6 – Gravelly Hill Interchange, better known as Spaghetti Junction – the most complicated motorway interchange in Britain, offering freeflow routes between the M6, A38M and A5127, linking various local roads and also allowing the passage of horse towlines, still in use at the time of construction in 1972. Most of this happens in three dimensions and is best seen from this canal, as roads snake through the air in all directions. It begins gently at first, a pipe bridge, a lattice canal footbridge and an arched cable bridge. Then it develops in earnest as slip roads flail through the air, the M6 crosses, the Birmingham to Lichfield railway passes through and ever more slip roads arc over, until the canal passes into a large box tunnel with two towpaths below the A38M Aston Expressway. It emerges into the relative simplicity of Salford Junction below the M6. The interchange deteriorated badly after its completion and repair work involved jacking up the deck of sections of slip road still in use to extract the decaying crosshead beams that needed to be replaced.

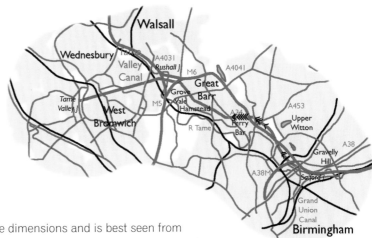

Part of the Gravelly Hill Interchange complex, through which the canal passes.

Birmingham & Fazeley Canal

The Birmingham & Fazeley Canal was designed as a connector linking a number of other canals. It leaves the New Main Line of the Birmingham Canal Navigations at Old Turn Junction. This was one of the most overcrowded points on the industrial canal network and, for this reason, lights were soon installed on the Farmer's Bridge lock flight, which went over to 24-hour working, rare in Britain.

The canal immediately passes Cambrian Wharf, a set of 18th century cottages and a toll office. The Flapper replaces former wharves and has one of the two wharf cranes incorporated into its outside seating area. Behind the Flapper is the 2013 Library of Birmingham.

Bridges in Birmingham generally have red doors in parapets to allow firemen to drop hoses through to the water source. This idea has been mirrored on a couple of recent bridges with railings, the railings having opening panels also painted red.

The Farmer's Bridge Flight, the Birmingham 13, is of interest to anyone with a liking for industrial and canal architecture. The top lock is one of a couple with tow rope guide pins. The brick towpath is ribbed to allow horses to grip.

Overflow weirs are of the bellmouth type, most discharging into side ponds below. Walkways are a feature of the canal and one promenade has been built upon a series of new brick arches. These extend an existing network of arches in

Farmer's Bridge Flight drops away from Cambrian Wharf.

two planes over a side pond. Owing to the lack of space, dividing barriers have been erected between the canal and the towpath below some of the locks. At lock 13, on the edge of the Jewellery Quarter, is a five-storey mural on a Mowlem office wall. It is best seen from the offside of the lock.

The dominant feature is Birmingham's BT Tower, which stands beside the canal. The concrete legs of the accompanying tower block stand in a lockside pond and all around the lock. It is a curious juxtaposition – the intricate detail of the locks contrasts with the grey slabs of concrete, the former well worn and mellow, the latter becoming chipped by passing boats.

In contrast, the huge brick cavern under Snow Hill station stands stark and empty but for some decorative ironwork. At one time the canal was heavily lined with dark brick walls, hemming it in and giving it the local name of the Bottom Road. These walls are now going, opening the towpath up as a pleasant place to walk.

Beyond the Aston Expressway and Corporation Street, Aston Junction allows the Digbeth Branch to peel off to the right to meet up with the Grand Union Canal. Aston Junction has been renovated in the best canal tradition. The flight is not without its interesting features, too, such as a recessed alcove in the brickwork of one bridge, the radius just large enough to take a lock balance beam and the height just enough for an adult to be able to stand.

Approaching Gravelly Hill Interchange of the M6, plant life begins to increase. The most common plant encountered is purple lupin, which adds a welcome splash of colour. Blackberry and elder bushes and reeds also help to bring the canal to life. Lupins have a foothold on the aqueduct over the River Tame, too. The canal now closely follows the river for most of its length.

Salford Junction, in the middle of Gravelly Hill Interchange, connects the Birmingham & Fazeley

Distance
33km from Old Turn Junction to Whittington Brook
Highlights
Farmer's Bridge Flight
Kingsbury Water Park
Navigation Authority
Canal & River Trust
Canal Society
Birmingham Canal Navigations Society
www.bcnsociety.co.uk
OS 1:50,000 Sheets
128 Derby & Burton upon Trent
139 Birmingham & Wolverhampton

with the Tame Valley Canal and the northern end of Britain's longest canal, the Grand Union. Although the locks are narrow, with single top and bottom gates, the canal itself is usually quite wide, no more so than where one factory completely spans the canal and towpath for 200m.

A factory that showed how to make a feature of the canal was the Cincinnati works, with lawns leading to the top of a bank covered with ivy, now empty. Beyond, the rusty fence style is revived.

The great cavern under Snow Hill station.

Unusually, the bridges on the Birmingham & Fazeley are named rather than numbered, and many of them are the traditional rounded arch type. One situated between the Boat and a steak bar displays a peculiarity that is to be seen regularly now – a small arched hole with a door leading back into one side of the bridge. This is for stop-logs, which can be dropped into the canal in the case of a breach.

The canal breaks into open country for the rest of its route, with views over farmland to the left. The square tower of the squat St Nicholas' church at Curdworth, with its finely carved font, can be seen over the hedgerow. Yellow irises become a regular feature, complemented later by the pink flowers of arrowhead.

Curdworth tunnel is 52m long. The towpath goes right the way through. It once carried notices asking boats to keep right, a fanciful idea as only the smallest boats would have much choice in the matter.

The top lock of the 11-lock Curdworth Flight has been rebuilt upstream out of the way of the M6 Toll. The A446 bridge just after the motorway shows off three different sizes of arch. The bridge has been progressively widened to take the present dual carriageway. Old canal buildings stand at the end of a track leading up to the Belfry Hotel and its golf course.

As Bodymoor Heath is approached, the noisy M42 moves alongside the canal before the waterway turns away through the 240ha Kingsbury Water Park, a set of former gravel pits. Brook Marston Farm Hotel may have a helicopter on the lawn next to the canal while the Dog & Doublet has a container of flowers over its lock bywash. The adjacent lane leads to Aston Villa football club's training ground.

The tank farm at Kingsbury's Oil Terminal can now be seen with higher land rising behind for the first time.

There are two locks beyond the Dog & Doublet and a swing bridge precedes the 38th and final lock on the canal. The swing bridge at Drayton Bassett is accompanied by a fixed footbridge that is quite unique to the British canal system. At each end is a Gothic arch doorway in a castellated brick tower with a spiral staircase, difficult for cyclists. Dosthill House is seen away to the right. Drayton Manor is unseen on the left. This was formerly the home of Sir Robert Peel, who was Prime Minister twice and invented the police.

Passing hosiery mills, the canal arrives at Fazeley Junction. This should have been the canal's end. John Smeaton completed it in 1789 but it had not been started until the owners of the Oxford, Coventry and Trent &

The unique bridge crossing the canal at Drayton Bassett, with castellated brick tower.

Autumn colours put on a splendid display in Hopwas Hays Wood near Tamworth.

Fazeley Junction with the Coventry Canal on the right.

Mersey Canals had all agreed to complete their own projects. Unfortunately, the Coventry Canal ran out of funds, even though it was to become one of the most profitable canals. The Trent & Mersey company built down as far as Whittington Brook, this section being purchased by the Coventry Canal, but the intervening section was completed by the Birmingham & Fazeley, this section following the contours for economy, rather than being straight to give speed.

The canal winds away along the left bank of the Tame valley, past willow-fringed meadows of grazing cows, heading towards the mast above Hopwas. In this village the canal is flanked by the Red Lion and the Tame Otter on opposite banks.

Hopwas Hays Wood carries notices that it is near firing ranges and the public are warned to keep out while red flags or lights are displayed. At this point the canal comes right alongside the river for a while, although it is not seen at the foot of the hillside. In turn, the river is replaced by the railway as a neighbour.

An Inland Waterways Association (IWA) plaque indicates the canal's change of ownership at Whittington Brook.

8 Coventry Canal

The Coventry Canal is an early contour canal that begins in the West Midlands and heads north-west across Warwickshire to the east of a ridge of higher land, into Staffordshire. It was promoted by Bedworth mine owners who planned to take Bedworth coal to the Trent & Mersey and Oxford Canals. Boats started shipping coal from 1769. It took a further 22 years to build the rest of the canal, during which time engineer James Brindley was sacked for spending too much of his time on other canal projects. There is still a statue of him in the basin at Coventry, though.

Once built, it was one of the most prosperous canals, originally carrying heavy coal traffic from the Warwickshire coalfield but also making much profit from selling water. It was still making a profit when it was nationalised in 1947, although narrowboat traffic ceased after WWII. As far as Lichfield, it is followed by the railway, initially the Coventry to Nuneaton link and then the West Coast Main Line. The Coventry Canal Art Trail also follows to Hawkesbury Junction.

Coventry takes its name from Cofa's tree, Cofa being a tribal leader c.600–650. Coventry's most famous resident, buried here, was Lady Godiva, who rode naked through the city as part of a challenge set by her husband, Loefric, Earl of Mercia,

Cash's Hundred Houses – a forward-thinking industrial and residential development.

to lower taxes in the 11th century. While there is a statue of her in the Broadgate, there are no fewer than three statues of the original Peeping Tom, the only person to look.

The 14th century St John's church was used as a prison when Cromwell defeated the Scots, hence the expression 'sent to Coventry'. The church of St Michael was formerly a priory, the bishop's seat 1095–1129, a parish church from 1373–1450 and a cathedral from 1918. Although the 90m steeple still stands, the rest was reduced to ruins in the 1940 Blitz that destroyed so much of Coventry in the longest single air raid on a British city during the Second World War. The new cathedral was designed by Sir Basil Spence and completed in 1962.

The Coventry Transport Museum houses the largest collection of British vehicles, spanning over 180 years of history, including Thrust 2 as well as the first safety cycle, which was built in Coventry in 1885 and followed the first penny farthing, built here in 1870.

The Bishop Street Basin of 1769 has two parallel arms, the western one being obstructed by a low swing bridge. The whole complex has been completely restored and is a well-signposted asset to the city.

The canal bends round to the left to face the Daimler Power House. Henry Lawson purchased a pair of cotton mills for the Daimler Motor

The old weighbridge office at Coventry.

Distance
43km from Coventry to Fazeley Junction and 9km from Whittington Brook to Fradley Junction
Highlights
Coventry Cathedral
Hawksbury Junction and steam pumping station
Oldbury Camp Iron Age hill fort, the reported site of Boudicca's defeat by the Romans
Navigation Authority
Canal & River Trust
Canal Society
Coventry Canal Society
www.covcanalsoc.org.uk
OS 1:50,000 Sheets
128 Derby & Burton upon Trent
139 Birmingham & Wolverhampton
140 Leicester

Company in 1896 and they became the Coventry Motor Mills. This company produced the first British car, the two-cylinder Daimler, which had tube ignition, solid tyres and tiller steering.

After Cash's Lane Bridge is a bank of impressive houses. These are Cash's Hundred Houses, although only 48 were actually built, of which 37 remain. They were owned by Joseph Cash, of woven nametapes fame. Built in 1857, workers' families lived on the first two floors and work was undertaken on the top floor. A lineshaft ran the length of the building and for 70 years provided power to assist individual workers to be competitive with factory looms.

Priestly's Bridge precedes the former Coventry Ordnance Works. When it was built in 1906 it was the largest workshop in Europe measuring 300m x 60m, and housed two huge lathes that made 380mm guns during the First World War. Two ferries operated across the canal for workers.

Hawkesbury Junction was built in 1802. It is also known as the Sutton Stop after the first lock keeper, whose family operated the lock from 1807 to 1876. There is a 150mm rise stop lock where the Oxford Canal enters. A southward-pointing spur shows the original proximity of the Oxford to the Coventry down to Longford.

Between the Second World War and 1970 this was the site of a Salvation Army mission for boat people at one of the busiest points on the canal network. The Greyhound had stables and is still popular. Now this is a landscaped conservation area and includes the Sowe Valley Footpath and Centenary Way.

The route becomes much busier as it forms part of a route from the Oxford Canal to the north-west, which avoids Birmingham's built-up areas, part of the Warwickshire Ring.

The covered dry dock at Hartshill Wharf.

The tall chimney just beyond the junction is on an 1837 steam pumphouse that raised water 35m from a well with one engine and from a stream underneath into the canal with another.

A series of lakes precedes a spur on the left before Bedworth that is the Newdigate Colliery arm, which includes a unique three-way lock.

The nearby Griff Hollows are said to be the original of the Red Deeps that featured in George Eliot's *The Mill on the Floss*, while the 18th century Gothic mansion of Arbury Hall became Cheveral Manor.

The Boot Inn stands at Boot Wharf. Nearby places of worship include a 1946 church built by German prisoners of war at Chilvers Coton,

the village appearing as Shepperton in George Eliot's *Scenes of Clerical Life* with Nuneaton being Milby.

Nuneaton had a priory and the 'nun ea tun' was the nun's river estate. With its redbrick houses it has been a mining town for 500 years, and a textile producer since the 1800s. The library has a collection of photographs and memorabilia relating to George Eliot, who was born at Arbury Hall in 1819.

The canal follows the River Anker to Tamworth. Hartshill Hayes Country Park stands on 55ha of wooded hillside, offering extensive views across the river valley.

The well-preserved Oldbury Camp Iron Age hill fort is the reported site of Boudicca's defeat by the Romans in 60 AD. The Romans were heavily outnumbered but killed 80,000 Britons. Boudicca poisoned herself to avoid capture.

One of the gems of the canal system is the 1840s Hartshill Wharf with a covered dry dock, restored blacksmith's shop in a waterways maintenance yard, refined clocktower with weathervane and a derrick crane. There is an air of Victorian Gothic with rounded windows, arches and curved walls to allow horses to pass with loads of wood for narrowboats. The wharf is an oasis in the countryside.

Atherstone claims the distinction of being 100 miles from London, Liverpool and Lincoln. On Shrove Tuesday, hundreds of people play medieval football in the streets. There are no rules and the shopkeepers take the precaution of boarding their windows.

Atherstone was the end of the Coventry Canal for a decade when the promoters ran out of money. In 1778 the Oxford Canal was opened from Banbury, exerting pressure for the Coventry Canal to be completed through to the Trent & Mersey Canal. The expensive but necessary flight of 11 locks lowered the canal by 25m. The locks are positioned close together at first but become more widely spaced lower down.

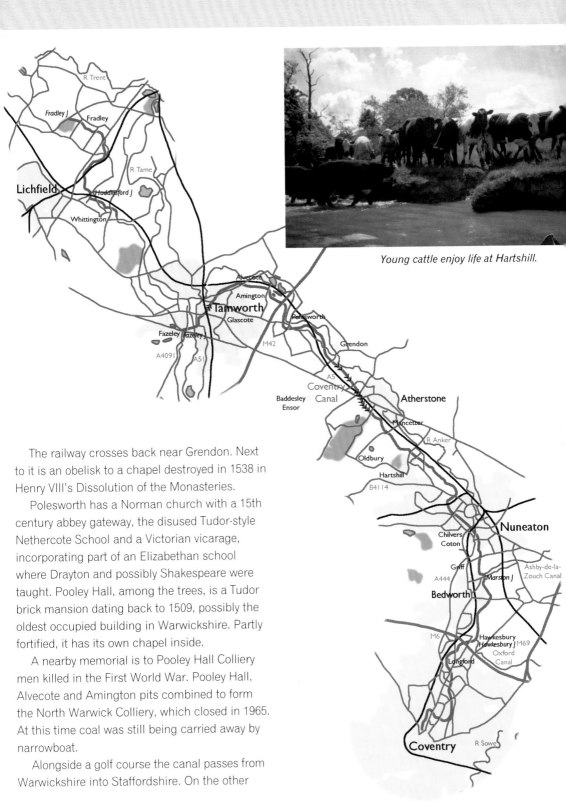

Young cattle enjoy life at Hartshill.

The railway crosses back near Grendon. Next to it is an obelisk to a chapel destroyed in 1538 in Henry VIII's Dissolution of the Monasteries.

Polesworth has a Norman church with a 15th century abbey gateway, the disused Tudor-style Nethercote School and a Victorian vicarage, incorporating part of an Elizabethan school where Drayton and possibly Shakespeare were taught. Pooley Hall, among the trees, is a Tudor brick mansion dating back to 1509, possibly the oldest occupied building in Warwickshire. Partly fortified, it has its own chapel inside.

A nearby memorial is to Pooley Hall Colliery men killed in the First World War. Pooley Hall, Alvecote and Amington pits combined to form the North Warwick Colliery, which closed in 1965. At this time coal was still being carried away by narrowboat.

Alongside a golf course the canal passes from Warwickshire into Staffordshire. On the other

The rural canal reaches away from the bottom lock of the flight at Grendon.

side of the canal are Alvecote Pools, flooded flashes resulting from mining subsidence. A nature trail has 100 species of bird and 250 flower species.

Housing follows the south side of the canal, together with the Gate Inn, until the west side also becomes built up at Bolehall. The Anchor, by a canal spur, precedes the final two locks at Glascote. The A5 has been diverted through here.

The derrick crane and the clocktower with its weathervane are just two of the features that give Hartshill maintenance yard its character.

The Burton upon Trent to Birmingham railway passes over and the A51 also crosses.

The first of two aqueducts, protected by a pill box, takes the canal over the River Tame, giving views down to Tamworth.

Tamworth, formerly Tomtun, was originally Saxon. Mostly grey stone, it is built at the confluence of the River Anker with the River Tame. It was razed to the ground by the Danes in 874 and again in 953. The 1180s castle has a rare Norman shell keep and tower, a medieval gatehouse, a Tudor chapel and a 15th century timber-framed great banqueting hall, Jacobean and Tudor apartments, Victorian suite and a bedroom haunted by the Black Lady, St Editha, the 9th century founder of Polesworth Abbey. The castle has been occupied for nearly 800 years. Offa, king of Mercia, had his capital and palace here in 757 and it was probably the stronghold of Ethelfleda, daughter of Alfred the Great.

The canal runs straight for 800m, to arrive at Fazeley Junction where it meets the Birmingham & Fazeley Canal.

The Coventry Canal reached Fazeley Junction (now restored) in 1790, by which time the

Birmingham & Fazeley Canal had already been extended to Whittington Brook. The Trent & Mersey Canal built the final section, then sold it to the Coventry Canal as a disconnected piece.

Huddlesford Junction used to bring in the Wyrley & Essington Canal, now just a spur as a result of being abandoned in 1954 but with restoration underway.

The Plough is tucked into a corner next to a minor road at Huddlesford, a small community cut in two by the Trent Valley Railway, which passes over for the last time.

To the left lies Lichfield. Between 1195 and 1310 the only medieval cathedral with three spires was built, the Ladies of the Vale, sited at the shrine of St Chad and built in local red sandstone. It has a 7th century manuscript of the gospels and a lady chapel with Flemish Herckenrode windows of 1802. It was badly damaged in the Civil War but the Victorian-restored west front has over a hundred carved figures.

The city was the birthplace of Samuel Johnson and Elias Ashmole, who left a collection of antiquities to Oxford University in the world's oldest museum, the Ashmolean.

Fradley developed because of an airfield, which stands beside the canal, its large hangars still dominating. The Coventry Canal joins the Trent & Mersey Canal at Fradley Junction, one of the busiest points on the canal system, the middle of a lock flight.

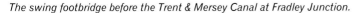

The swing footbridge before the Trent & Mersey Canal at Fradley Junction.

Ashby-de-la-Zouch Canal

The Ashby-de-la-Zouch Canal, Ashby Canal or Moira Cut was intended to be a major link in our canal system, extending to the Trent valley to join the Commercial Canal and then serving to Liverpool or Manchester. Running north across Warwickshire and Leicestershire from the Coventry Canal, it never got any further north than Overseal and never even reached Ashby-de-la-Zouch itself. To get down to the Trent would have required heavy and expensive locking. The canal, as it remains, has no locks at all except at Moira on the disconnected section, and it is part of one of the longest level pounds in Britain as it follows the 91m contour. Extensions northwards were as plateways or tramroads and the canal carried heavy coal traffic, although it was not very profitable. In 1804, the year it was opened, good-quality coal was found at Moira, supplied by canal to Oxford colleges, among other customers. Medicinal springs were found at the canal head in the 19th century, resulting in passenger boat traffic, and the canal company also tried exporting cheese. Something they have successfully exported is Measham ware, Church Gresley's highly prized decorated brown canal earthenware.

The canal was bought out in 1846 by the London, Midland and Scottish (LMS) railway. In 1856 there were experiments with the steam-powered barges *Pioneer* and *Volunteer*. Traffic peaked at 153,000t in 1870. In 1918 there was extensive mining subsidence in the Measham area and by the 1940s much of the canal around Donisthorpe was on embankment or low bridges. The LMS offered to give the canal to the Coventry Canal Company but they declined. In 1944 the canal was shortened to Donisthorpe, abandoned north of Measham in 1957 and shortened again to its present terminus north of Snarestone in 1963. The last regular coal carriage from the North Leicestershire Coalfield took place in 1970 by the Ashby Canal Association, which was founded by an angling club and managed commercial carrying as far as Brentford in order to fund it. The closures not only avoided the cost of repairs but also allowed further extraction of coal from beneath the line of the canal.

There are plans to reopen the canal to Measham, the Ashby Canal Association owning the canal bed as far as Gilwiskan Aqueduct. Around Moira Furnace in Donisthorpe Country Park 2km has been restored. Full restoration is intended.

Farmland is rarely more than gently undulating and views are frequently extensive,

Distance
39km from Marston Junction to Bath Yard Basin

Highlights
Battlefield Line heritage railway
Site of the Battle of Bosworth Field, 1485
Moira Furnace
Conkers Waterside Centre

Navigation Authority
Canal & River Trust

Canal Society
Ashby Canal Association
www.ashbycanal.org.uk

OS 1:50,000 Sheets
128 Derby & Burton upon Trent
140 Leicester

The furnace at Moira – the main reason for the canal being built to this point.

The pumping station at the present foot of the main section of the canal.

hedges often being absent beside the canal. Views are better from the water before the reedmace grows too much in the summer. Bridges across the canal are mostly arches of stone or blue engineering brick, sometimes clearly distressed by mining settlement.

These days the canal is almost entirely rural and, being a dead end, is not heavily used by canal traffic although its lack of locks and its rural feel entice enough boats up to keep the water stirred up.

Although the Ashby-de-la-Zouch Canal is, for the most part, on one level, the lock-free opportunities continue from Coventry or from Atherstone. The canal leaves the Coventry Canal

Looking along the line from the Coventry Canal at Marston Junction.

at Marston Junction on the edge of the old mining town of Bedworth. A stop lock was built here to control traffic. The siting of the junction is the result of a late change of plan, the original intention being to join at a Griff Colliery arm to the north.

The first two bridges are Grade II structures made of Attleborough sandstone. At Marston Jabbett the canal is also crossed by the Trent Valley Railway line. Three different groups of powerlines pass over the initial reaches of the canal, diverging from Hawkesbury.

After the bridge at Whitestone, a collection of chalet buildings stand on top of the steep canal embankment. One of the owners has laid out his section of bank as a shrine to Liverpool Football Club.

Bramcote's hospital is next to the canal and to the B4114, where vehicles are encouraged to slow to the point where they don't become airborne from the canal bridge.

On the hillside is Burton Hastings, where the font of St Botulph's church dates from the year 1300. Two limestone bridges across the canal are also Grade II structures. A distinctive powerline flyover beside the cut, with lines along and across the canal, is hard to ignore. The hillside, where Stretton Baskerville Village was lost to the field enclosures of the 16th century, now serves to eliminate any noise from the M69, which is 900m away.

Next to the Lime Kilns public house, the A5, built as the Roman Watling Street, passes overhead and the canal moves from Warwickshire into Leicestershire.

The A47 feeds the only built-up part of the canal, Hinckley. The centre of the town is marked with a transmission mast that dwarfs the church spire. Hinckley has some fine timber-framed cottages and its industrial background includes having the stocking frame in 1640, long before Leicester did. It was also the home of architect Joseph Hansom, designer of the Hansom Cab, and of Triumph motorcycles.

Powerlines cross the canal near Higham Grange hospital and kennels make their presence heard. An ancient fishpond site follows a field of Shetland ponies.

The line of the branch railway from Hinckley runs beside the canal. Although the branch was laid with track and signalled, it never carried traffic and was dismantled in 1900. It is joined by the route of the railway from Nuneaton which approaches past the MIRA test track complex.

From the picnic area at Stoke Golding, signs point to the George & Dragon, while a striking spire draws attention to the 13th century church of St Margaret with its decorated arcade and delicate window tracery. The village is known for the Stoke Golding Country Dance, a more complex version of the Scottish Strip the Willow.

Dadlington's church dates from the 13th century but the Dog & Hedgehog, with its green and yellow illuminated signs, looks horribly modern. A narrowboat appears to be parked in a field but there is a slot cut at a right angle to the canal just large enough to hold it.

To the west of the canal is the site of one of the turning points in British history. In August 1485, Richard III was killed at the Battle of Bosworth Field and Henry Tudor was crowned on the battlefield as Henry VII, the first Tudor king, thus ending 30 years of the Wars of the Roses. It was the last time the British crown was to change hands on a battlefield.

Many other narrowboats are moored along the canal. The canal crosses the line of the Roman road from Mancetter to Leicester before Sutton Wharf Bridge. Tea rooms and trip boats at the wharf make this a busy area.

These days there is a visitor centre with battle re-enactments, jousting, Morris and tea dancing and falconry among the attractions at various times. There is a birdwatching hide at the edge of the wood, looking down towards the canal, which is edged with gorse bushes here.

The current terminus of the Battlefield Line is at Shenton, which has been in the process of being restored since 1970. It uses ex-industrial steam and diesel locomotives. The abandoned route of the railway could see future extension, following the line of the canal. The ticket office was built in Leicester but moved to Shenton in 1993 as part of a road-widening scheme.

Near the terminus is a very amateur-looking jetty but all its wooden handrail posts are topped with beautifully turned wooden knobs. A high brick aqueduct carries the canal over the road from Shenton, which has a Victorian church with a 17th century monument to the Battle of Bosworth Field. Shenton Hall dates from 1629 and was much rebuilt in the 19th century.

Market Bosworth has some fine thatched cottages. It was where Dr Johnson taught for a time. A timber yard is located next to a road bridge. Dredging in 1976 brought up a First World War bomb in dangerous condition, which had been dropped from an airship and was intended for the local gasworks. The canal society members placed mileposts along the canal and one here shows it will be the midpoint of the canal when it has been fully restored.

Rooks circle above their rookery in a wood in the spring. A windsock on the outskirts of the town marks a landing strip. Beyond is the town itself with the slender spire of the partly 14th century church visible.

From Carlton Bridge, where it passes over the Sence Brook, the canal dances past Congerstone with its kingfishers, wrens and water voles.

The aqueduct and listed bridge over the River Sence at Shackerstone station.

A distinctive brick tower rises above Barton in the Beans. There is a railway museum at Shackerstone. A Grade II viaduct with delicate cast-iron parapets carries a road over from the station. The canal follows the River Sence, with its otters, before turning sharply to cross it on a stone aqueduct.

A line of narrowboat moorings curves round a medieval pond in front of the village, which dates from Saxon times and from which Saxon jewellery has been unearthed. Prominent signing makes the Rising Sun a difficult public house to miss, should you wish to visit it.

Some stands of woodland are found alongside this section of canal: oaks, hawthorns, willows and, appropriately, ashes, with coltsfoot and butterbur among them along the banks in the spring. The woods hide Gopsall Park, the grounds of the former Gopsall House where Handel may have composed the *Messiah*.

A picnic table at Gopsall Wharf is one of a number that have been placed along the length of the canal.

A rabbit warren has been dug in the embankment between the canal and the line of the former railway, which used to cross to the south of Snarestone.

The canal's major structure is the 230m long tunnel at Snarestone. It is unusual for a canal to end beyond a tunnel rather than before a collapsed bore. The line of the canal snakes through the tunnel but the full bore of the tunnel is visible all the way through. As the tunnel has no towpath, the line of the former horse path leads from before the Globe Inn over the top of the low hill, on which the 18th century farming village of Snarestone is built.

The canal currently ends just above a Victorian Gothic pumping station, built in 1892 to supply water to Hinckley, its chimney conspicuous in the local countryside. The canal's dead end makes it susceptible to collecting floating debris and an oil film. The nearest access point is a gate, kept locked by the canal society.

The present line ends just outside the National Forest and is mostly in a Conservation Area that boasts nine species of dragonfly (including the rare red-eyed damsel), the flat-stalked pondweed (which is rare elsewhere) and a fair sample of herons, coots, moorhens, mallards, kingfishers and a few swans. Restoration to Donisthorpe will make use of a disused railway line where the canal line is no longer accessible. Measham station has become Measham Museum.

Beginning at Donisthorpe is a 2km section of canal that is wide, fully restored and in excellent condition, home to a canal society trip boat and a maintenance craft.

Initially it runs between a minor road and woodland planted on colliery spoil. After limekilns it is crossed by an arched bridge that feeds directly into Moira Furnace, which is now partially restored as a museum. It was an 1805 blast furnace manufacturing pig-iron with the compressed air blast produced by beam engine, but was closed after two years since it was uneconomical.

Beyond a swing bridge the canal passes teasels and a coach depot as it swings into the back of Moira, named after Moira in Northern Ireland. The tower of a fire station locates Moira Lock, installed because of 3m of subsidence.

Beyond the road crossing there is low ground on the right towards Sarah's Wood. The canal turns under the Marquis footbridge, named after the local colliery, supported by a pyramid of poles, to enter the terminus at Bath Yard Basin, also named after a colliery, one of 28 in the South Derbyshire Coalfield by the canal at various times. Conkers Waterside has room for future moorings but is presently more appreciated by families using the slides and other facilities. The canal stops just short of the Leicester to Burton upon Trent railway and 8km from its intended target of the River Trent.

Snarestone tunnel is short and the bends are not obvious.

10 Oxford Canal

The Oxford canal formed part of the original Grand Cross scheme to link the Thames, Mersey, Trent and Severn. From 1769 until the time of his death in 1772, James Brindley took the Oxford Canal from Coventry as far as Brinklow. Samuel Simcock took over, reaching Banbury in 1778, then Robert Whitworth continued to the River Thames by 1790. A contour canal, it had many meanders, especially on the northern section. It lost traffic after 1805 when the Grand Union Canal provided competition. In response, it was shortened between 1828 and 1834, cutting 22km off its length. From 1840, it lost further traffic to the railways. Even so, it remained profitable and paid a dividend right up to nationalisation in 1947. The last commercial traffic was carried in the 1950s but it is now one of the busiest canals in the country.

The Oxford Canal leaves the Coventry Canal at Hawkesbury Junction. It is approached under a cast-iron bridge, one of a number of these graceful structures, particularly on the northern section.

Looking back to Cropredy.

The canal then moves from Warwickshire into the West Midlands as far as Sowe Common. The Wyken Old Colliery Branch has been provided with convenient moorings by the Coventry Canal Society. At Sowe Common there used to be a cast-iron towpath bridge but it was removed during the construction of the M6. It now crosses the River Sherbourne at Spon End in Coventry.

An 18th century Georgian hall is now a hotel hidden in the trees next to a small church with a delicate spire. Also in the vicinity is the Rose & Castle, a public house with a suitably canal-inspired name. There is a medieval ridge and furrow system, the first of a number that have survived along the line of this canal.

There are extensive views to the south-east towards Nettle Hill. Where once the canal made a loop up the Hopsford valley, it now crosses on an aqueduct. A riveted viaduct passes high over the canal and railway, its sides supported with scrolled stays. The canal heads towards the Upper Smite Village site. The M6 also passes over and heads east.

Stretton Stop lock and toll office are approached by a plank footbridge, often swung across the water but easily swung back. The Stretton Wharf Branch goes north and is used by

Distance

37km from Hawkesbury Junction to Braunston Turn and 81km from Napton to Oxford

Highlights

Extreme contouring near Wormleighton

Tooley's Boatyard, Banbury

Colleges of the University of Oxford

Navigation Authority

Canal & River Trust

Canal Society

Friends of Oxford Canal & Basin

OS 1:50,000 Sheets

140 Leicester

151 Stratford-upon-Avon

152 Northampton & Milton Keynes

164 Oxford

Rose Narrowboats, based here beside the Fosse Way Roman road. At Brinklow there is a motte and bailey, built in the time of King Stephen to defend the road.

The canal was barred from entering the park at Newbold Revell by the landowner so the 12 Brinklow Arches were built over the Smite Brook. Some were used as stables, forges, hay stores and dwellings although 11 were filled in when the canal was widened in 1834.

The stop lock and roving bridge at Hawkesbury Junction.

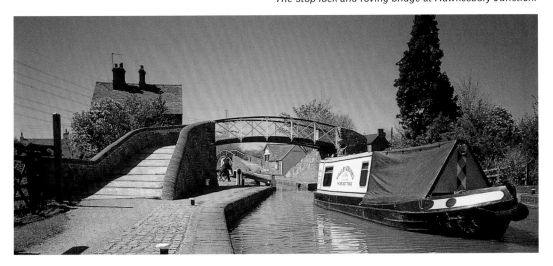

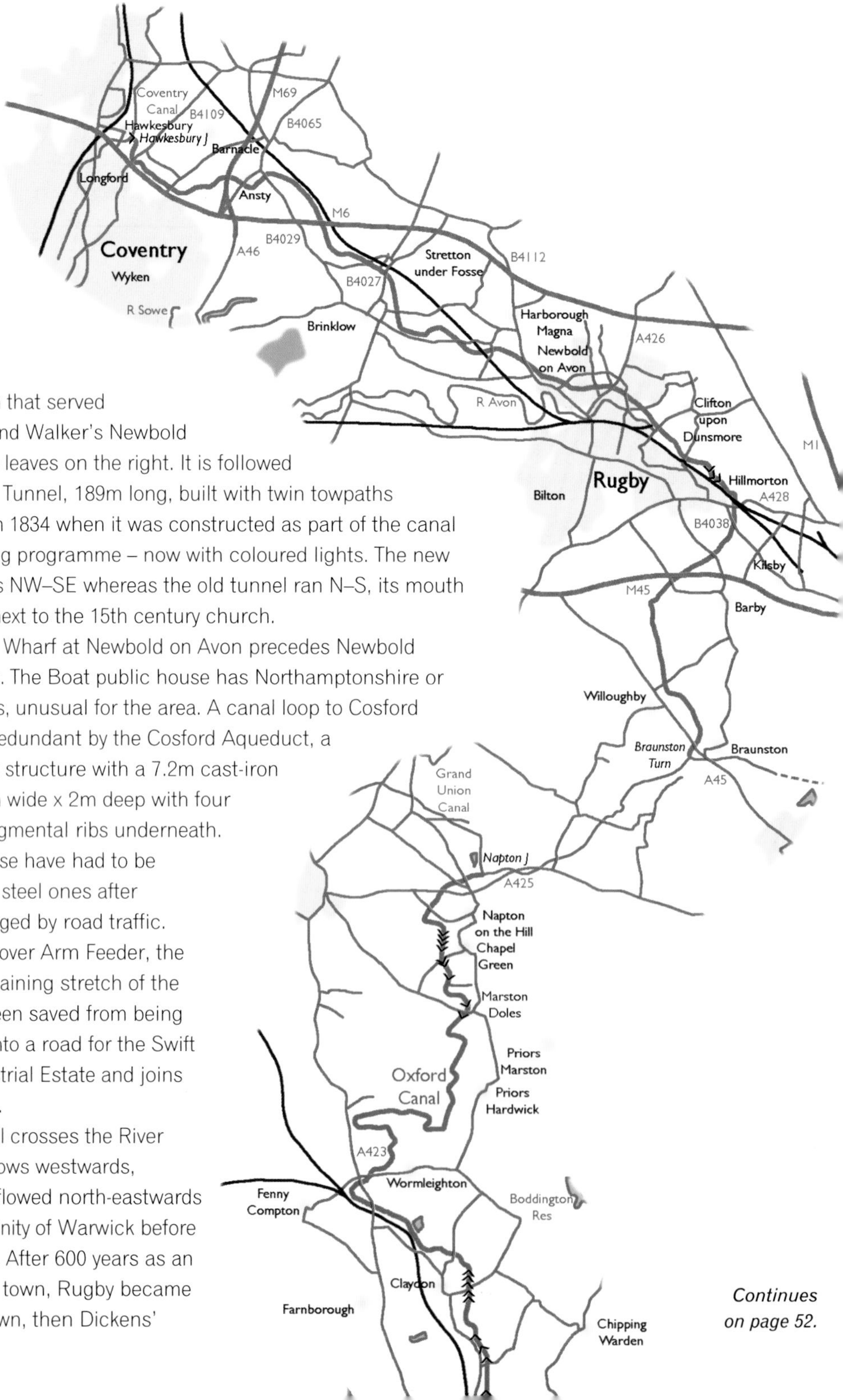

Coventry Canal
B4109
Hawkesbury J
Hawkesbury J
Barnacle
M69
B4065
Longford
Ansty
M6
Coventry
Wyken
A46
B4029
B4027
Stretton under Fosse
B4112
R Sowe
Brinklow
Harborough Magna
Newbold on Avon
A426

A branch that served
Norman's and Walker's Newbold
Lime Works leaves on the right. It is followed
by Newbold Tunnel, 189m long, built with twin towpaths
– unusual in 1834 when it was constructed as part of the canal
straightening programme – now with coloured lights. The new
alignment is NW–SE whereas the old tunnel ran N–S, its mouth
remaining next to the 15th century church.

Newbold Wharf at Newbold on Avon precedes Newbold
Quarry Park. The Boat public house has Northamptonshire or
table skittles, unusual for the area. A canal loop to Cosford
was made redundant by the Cosford Aqueduct, a
magnificent structure with a 7.2m cast-iron
trough 4.6m wide x 2m deep with four
cast-iron segmental ribs underneath.
Three of these have had to be
replaced by steel ones after
being damaged by road traffic.
The Brownsover Arm Feeder, the
longest remaining stretch of the
loop, has been saved from being
converted into a road for the Swift
Valley Industrial Estate and joins
by the A426.

The canal crosses the River
Avon that flows westwards,
although it flowed north-eastwards
from the vicinity of Warwick before
the Ice Age. After 600 years as an
agricultural town, Rugby became
a railway town, then Dickens'

R. Avon
Bilton
Rugby
Clifton upon Dunsmore
M1
Hillmorton
A428
B4038
Kilsby
M45
Barby
Willoughby
Braunston Turn
Braunston
A45
Grand Union Canal
Napton J
A425
Napton on the Hill
Chapel Green
Marston Doles
Priors Marston
Oxford Canal
Priors Hardwick
A423
Wormleighton
Boddington Res
Fenny Compton
Claydon
Farnborough
Chipping Warden

Continues
on page 52.

Mugby Junction, before turning to heavy electrical engineering. The name has become synonymous with rugby football, first played at Rugby school in 1823; the school featured in *Tom Brown's Schooldays*. Rupert Brooke was born here. Another building of note is the 14th century St Andrew's church with its 1879 nave and tower by Butterfield, probably the only church in England with a double peal of bells.

Hillmorton has three locks to begin the 23m 13-lock rise to Marston Doles. In 1840 these three were doubled with connections between the pairs so that they could act as side ponds for each other. Unusually, the pairs remain operational but unconnected and pleasure boat use is so heavy that they are needed.

Dominating the area was the Rugby Radio Station, established in 1924 on a 3.6km² site, transmitting around the world. The tallest masts weighed 200t and were 250m high, the world's tallest structures when erected. The last tall mast was demolished in 2007. At this point the canal changes from its south-easterly course to head south-west.

Braunston is one of the best-known village names on the canal system although most of the activity is to the east on the Grand Union Canal. The approach from the north is quiet enough, herons and swans minding their own businesses below a ridge topped by a disused windmill and a church. This has a tall Victorian spire and is in a churchyard that includes many boatmen's graves. The Oxford Canal originally continued south-west but this section was taken over as part of the Grand Union Canal main line and southbound boats now traverse this stretch in the opposite direction from when it was first built by the Oxford Canal Company.

Incorporation of the Oxford section and the 800m Braunston Branch saved £50,000 in construction costs along the Grand Union Canal route from Birmingham to London. An idea of the importance of this new transport link can be gauged from the fact that goods being conveyed from London to Abingdon were sent via Braunston rather than up the River Thames because of the poor state of the latter.

The southern section of the Oxford Canal saw horsedrawn boats long after other canals had given them up and there can be few long

Running through All Oaks Wood.

sections of canal better suited to the pace of the horse. Joseph Skinner's mule-drawn boat was in use until 1959 and is now in the National Waterways Museum, Ellesmere Port. The rural feel is complemented by bollards made from roughly hewn tree trunks. The southern section of the canal has many angling contests.

Napton on the Hill takes its name from the Anglo-Saxon *cnaepp tun*, meaning hilltop homestead, the hill being over 120m high. On top is a windmill, restored in the 1800s, although there has been a mill here since at least 1543. Materials for building the 13th century St Lawrence's church were left by the green, ready to start work, but overnight they reappeared at the top of the hill so it was built up there.

The Olde King's Head is just up from Napton Wharf, the base for Napton Narrowboats. By Napton Bottom Lock is a farmhouse that served beer to boat crews, becoming the Bull & Butcher and now the Folly, noted for its selection of pies. A notice warns that it is the last pub for five hours. There are ten locks to be negotiated in the next 3km up to Marston Doles.

The top lock is by the Welsh Road at Marston Doles. Picnic tables wait by the lock side and the 16th century Holly Bush Inn in Priors Marston offers a transport service to and from the lock.

The summit pound runs for 18km and is the most extreme example of a contour canal. It is a rural area with very few buildings along its length but the canal seems to go out of its way to visit all of them and search for more as well. It suffers from water shortages in the summer but a back-pumping scheme has been installed to get round the problems and flows have been augmented with mine water from Wolverhampton.

Parts of the stone church in Priors Hardwick date from the 13th century. The village was badly hit by the Black Death and many of the houses were pulled down by Cistercian monks in the 14th century. Along the road, Stoneton Village has disappeared almost without trace; Stoneton

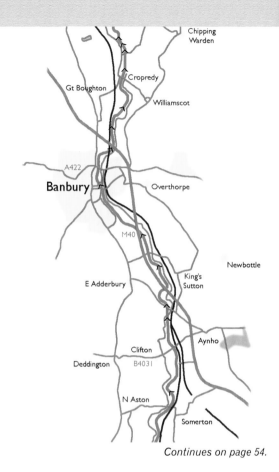

Continues on page 54.

Moat Farm is one remaining link with the past.

The canal uses a gap cut by overflow from an Ice Age glacial lake between Leicester and Stratford-upon-Avon. The next kilometre is called the Tunnel because it was indeed a tunnel until it was opened out in 1868. It is now a steep-sided cutting.

The boundary between Warwickshire and Oxfordshire comes directly after the first of the lifting bridges that are so distinctive of this canal, built with heavy balance beams but no superstructures. They are particularly to be found south of Banbury and are usually left open.

Five locks at Claydon start the 30-lock fall through 59m to the River Thames with old stables at the second lock.

The River Cherwell now runs alongside and follows closely to Thrupp. There was a bridge

The twinned Hillmorton locks.

across the river by 1314, but it was the bridge in place in June 1644 that was the focal point for the Civil War Battle of Cropredy. Cromwell's forces, under Waller, attacked a smaller Royalist group in an attempt to force a passage towards Oxford. They were beaten back and lost their artillery. Some of the soldiers were buried in the churchyard of the sandstone church of St Mary the Virgin, which has fine woodwork. It also has a mark on the vestry floor that may be the bloodstain from a young messenger killed there after the battle. The brass eagle lectern spent 50 years in the river after being hidden there from the troops by the villagers. The footsteps of a man, woman and girl killed by a Roundhead are sometimes heard at the 15th century Red Lion.

The Old Canal Wharf used to be a toll collection point. These days people pay for the afternoon teas it provides.

Banbury begins with a foundry that takes its cooling water from the canal. A westerly wind brings the aromas of Maxwell House coffee from Britain's largest coffee factory, cocoa, Bird's custard and baking bread from the Fine Lady ovens.

Banbury has had a poor record of civic vandalism in the past. The cross of nursery rhyme fame was pulled down by the Puritans in 1602, the fine lady probably being a girl in a May Day procession. The current cross is a replica of 1859. The Perpendicular Cathedral of North Oxfordshire needed renovation; instead it was blown up (a cheaper option), the 'unsafe' portion surviving the explosion. Its replacement of 1793 by SP Cockrell is described as looking like a gaol.

Banbury, from the Old English for Banna's stronghold, was a wool town and formerly had the largest cattle market in Europe. It has not always been as idyllic as it seems, however. Above the ridge on the left is the Upper Heyford Airfield, one of the two bases from which the US Air Force bombed Tripoli. Now closed, it was one of the three finalists for the national sporting academy venue and the British Olympic Association's preferred choice, but it lost out to Sheffield.

Hanging judge Sir Francis Page, who lived from 1661 to 1741, sold Middle Aston to Sir Clement Cottrell Dormer and concentrated on Steeple Aston. Here he had graves in St Peter's church moved to make way for a large one for himself and his wife. The 1730 Page monument shows the 70-year-old judge seated in his robes with his 40-year-old wife lying rather less robed at his feet.

The railway crosses at a point where nine carriages fell on to the frozen canal on Christmas Eve 1874 in the Great Western Railway (GWR)'s worst accident, killing 34 passengers and injuring 65 more. The 15-coach

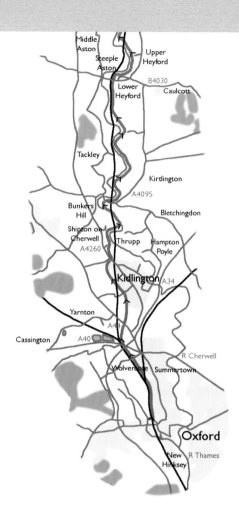

train was carrying 500 passengers behind a pair of single driving wheel engines when a wheel on the leading coach broke, with disastrous consequences. Amazingly, the Oxford stationmaster had a rescue train underway just half an hour after receiving a telegram about the accident.

At Thrupp Wide the canal again uses the course of the river but this time the river has been moved to a new channel on the east side. Thrupp has been described as the perfect canal village; cottages stand back from the grassed canal embankment. The Boat Inn provides refreshment, as does the Jolly Boatman.

The bridge by Bourton House.

The bottom of the Napton Flight with the windmill above.

The graceful cast-iron roving bridge in the Tunnel.

The Plough pub is passed at Upper Wolvercote and the Marylebone to Oxford line is the final railway crossing at Summertown, where the playing fields of St Edward's School occupy the left bank. Pupils have included Kenneth Grahame, whose views of the canal may have been included in his *The Wind in the Willows*.

The canal here and its boats and crews were central to the early part of *Northern Lights*, the author of which, local resident Philip Pullman, was among those demonstrating against a housing development on the Castle Mill boatyard site. PD James was also born in the city and set *The Children of Men* here, including passing reference to the canal. A chain ferry punt used to cross here and there was a boatmen's chapel mounted on a barge hull in the parallel Castle Mill Stream in the 19th century. Behind is the Oxford University Press. The university has been publishing for over 500 years and is Britain's most prolific imprint, although not the most efficient as it took them 35 years to get back the corrected proofs for *Coventry Constables' Presentments*.

Looking down the canal with the spire of Kings Sutton church rising out of the mist.

The canal drops down through Louse or Isis Lock, originally built wide beam but now narrowed, below an ornate iron roving bridge of 1796. At a brick headwall there was a further lock to the river, Armchair Weir, but it had only a single gate and was wasteful of water. Crossing Castle Mill Stream, it is possible to proceed up the Sheepwash Channel to the River Thames or Isis opposite the Tumbling Bay bathing place. This route formed part of the improved Liverpool to London link as the Thames was in poor condition upstream to the Duke's Cut. A local improvement was the decision to fix a swing railway bridge in the open position in 1984. Until then, railway crews had to unbolt the track every time a boat wanted to pass. In 1995 the station, which features on the opening page of *Zuleika Dobson*, had the proud boast that improved security had reduced the number of thefts from its car park over the year from 46 to a mere 16.

Oxford is usually taken to be the ford where oxen crossed but it may be significant that the Old English for salmon was *ehoc*. Confusingly, the River Ock is downstream at Abingdon.

Over 2,000 years ago, Lud had the realm of southern England measured and discovered that Oxford was at the exact centre. St Frideswide built a monastery in 727 and began the walled town. In 912 Oxford was used by Edward the Elder as a buffer between Wessex and the invading Danes. It was the sixth-largest town in England in 1066 and it was still the fifth-largest in 1781. Oxford Castle was built in 1071 for William the Conqueror by Robert d'Oilly and includes a Saxon stone tower. Empress Matilda escaped from the castle and King Stephen's seige by crossing the frozen Thames dressed in white one snowy night in 1142. It was Charles I's headquarters in the Civil War and Cromwell and Hitler both planned to make it their capital. In 1996 the prison was converted to a more welcoming hotel and shopping centre.

Unusual towpath viaduct passes over as the canal crosses the River Cherwell on the level above Aynho Weir.

The university was in existence by the 11th century, the oldest in the English speaking world. The oldest of the 35 colleges dates from 1249 while the youngest don was 17-year-old Henry Phillpotts at Magdalen in 1795. The colleges were heated by Moira coal, brought via the Ashby-de-la-Zouch Canal. Defoe's Moll Flanders and an extravagant husband blew £93 in a 12 day visit to the city and it was visited by Frankenstein. Evelyn Waugh's *Brideshead Revisited* saw much of its early action in the colleges, Adam Bell was a fellow in Elizabeth Gaskell's *North and South* and, in the same way, much of the scene-setting for Ian McEwan's *On Chesil Beach* took place around north Oxford. Fictitious students include Tibby Schlegel in EM Forster's *Howards End*, Scott's Guy Mannering, Tom Brown and Jim Brading in Arthur Ransome's *We Didn't Mean to Go to Sea*.

The city has been used for filming *Inspector Morse*, *A Fish Called Wanda*, *The Madness of King George*, *Shadowlands* and *Waiting for God*. To Matthew Arnold it was the 'home of lost causes' and it was where Lewis Carroll composed *Alice's Adventures in Wonderland*. JRR Tolkien and CS Lewis were resident writers and astronomer Edmond Halley was also here.

Oxford buildings of note include Blackwell's bookshop, the Bodleian Library and the world's oldest museum, the Ashmolean, which dates from the 1840s although the collection was only begun in 1683.

Central Oxford with its colleges.

Grand Union Canal

The Grand Union Canal is the backbone of the canal system – Britain's longest canal, joining Britain's two largest cities – yet it was not named as such until 1929. It began life as at least eight separate canals with William Jessop doing the major building work between 1793 and 1805. He completed the line, which avoided most of the winding Oxford Canal and the Thames in poor condition, shortening the route to London by 100km, although there were many locks. During the Second World War it became an important route for transporting Birmingham arms equipment and coal to London, often by teams of three female workers who, from their Inland Waterways (IW) badges, were nicknamed Idle Women. Freight carriage did not end until about 1970, later than on most canals, although there was a marked decline after the severe winter of 1963.

The start was originally at the Warwick Bar, where it met the Digbeth Branch. In order to relieve congestion in the city centre, the Birmingham & Warwick Junction Canal from Bordesley Junction was opened in 1844 to meet the Birmingham & Fazeley Canal at Salford Junction. The Tame Valley Canal was also built to the same point and opened at the same time. These days the sky above the junction is dominated by Junction 6 of the M6 – Gravelly Hill Interchange or Spaghetti Junction – as spurs loop in all directions. It is a long time since this was the Old English *Beormings ham*, the village of Beorma's people.

The imposing church overlooking Camphill second lock.

The top lock raises the canal to the 16km long 117m level, the first of the canal's summit levels. The top lock is also the last of the narrow beam ones. Once the smaller canals were amalgamated it was decided to widen the whole route to broad beam in the 1930s as a work creation scheme during the Depression. However, the final locks into the city were never converted.

Smoke from the opposite side of the canal may be from the Birmingham Railway Museum, which is based in a former Great Western Railway shed and has live steam with engines; wagons; coaches; the royal carriage of Edward VII and Queen Alexandra; the royal carriage used by Churchill and Eisenhower as an office during the Second World War; and a travelling post office that was attacked in the Great Train Robbery.

There are moorings between the B4101 and the top of the Knowle lock flight but it seems that many powered craft do not bother coming up any further than the foot of the flight, resulting in increased traffic below. The locks, as often elsewhere, are conveniently grouped together. Five broad-beam locks stand beside the six narrow ones they replaced, covered by concrete slabs. The new ones have distinctive 1930s Ham Baker candlestick paddle gear, tall white cylinders leaning slightly outwards, seen from here to Napton Junction. Around them is an assortment of side ponds and flower beds planted with pansies, the first side pond at an oblique angle. Strangely, the new concrete locks are above ground level rather than being recessed where the ground could help resist the pressure of the water on the lock walls.

The A4141 crosses King's Arms Bridge. The adjacent public house is the Heron's Nest. Close by is Heronfield Ark animal sanctuary. Some of the sheep near the canal have more than their fair share of horns. The Black Boy offers an alternative canalside watering hole. The West Midlands boundary follows the canal for a kilometre before giving way to Warwickshire.

Distance
215km from Salford Junction to the River Thames

Highlights
Birmingham Railway Museum, featuring Churchill and Eisenhower's Second World War office train and the travelling post office attacked in the Great Train Robbery
Hatton lock flight, popularly known as the 'Stairway to Heaven' – the best locks on the Grand Union Canal
Warwick Castle, England's finest medieval castle
Braunston Turn and Marina
Stoke Bruerne Waterways Museum, Britain's first canal museum
Chilton Summit Locks
Syon House and its grounds, steeped in 1,000 years of royal history
Royal Botanic Gardens at Kew, the world's largest collection of living plants

Navigation Authority
Canal & River Trust

Canal Society
Braunston Canal Society
www.braunstoncanalsociety.org.uk

OS 1:50,000 Sheets
139 Birmingham & Wolverhampton
151 Stratford-upon-Avon
152 Northampton & Milton Keynes
165 Aylesbury & Leighton Buzzard
166 Luton & Hertford
176 West London

One of the most notable houses in the area is the moated medieval brick and stone manor of Baddesley Clinton. It was a haven for persecuted Catholics and has three priest holes where nine men stood ankle deep in water for many hours in 1591 to evade capture. Little has changed since 1634, except for the 1920–1930s equipment in the medieval kitchen.

An immaculately manicured hamlet at Turner's Green includes the Tom o' the Wood public house, named after a former windmill.

The motorway is kept away from the canal by a triangular railway junction that used to lead to Cheltenham but now runs out of steam at Stratford. For reasons that soon become apparent, there are fine views eastwards on both sides of the canal.

The Stairway to Heaven is the popular name for the Hatton lock flight, the best locks on the Grand Union Canal: 21 locks drop the canal 45m over 3.1km. Former stables near the top lock house a canal shop, while the Hatton Arms offers stronger liquid refreshment with a fine view from above the fourth lock.

Warwick Castle, founded in 915 by Ethelfleda and developed in 1068 by William the Conqueror, is said to be the finest medieval castle in England. The name comes from the Old English *wering wic* meaning workplace or trading centre at the weir. To the north, on the Avon at Guy's Cliffe, is a cave carved by Guy of Warwick in 929, where he lived as a hermit after fighting the Danes and then being rejected in love.

Royal Leamington Spa was noted for its saltwater springs from 1586 but did not get its royal charter from Queen Victoria until 1838. It has Georgian and Victorian housing, is the home of lawn tennis and hosts the women's national bowls championships each summer, a touch of elegance not seen from the canal. Near the station is the art gallery and museum in the Royal Pump Rooms of 1814, which were built over one of the town's seven springs. Jephson Gardens, with their glasshouse, were named after the local Dr Jephson who, in the 19th century, developed the spa water concept and was responsible for the town's medical reputation and mini Eden.

Roman coins have been found at Radford Semele, the reed ford of Norman landowner Henry de Simely in the days of Henry I. Flints and axes show the settlement to be at least 30,000 years old. The hilltop church is now Victorian although its appearance seems to have more in common with its 1100 rebuild. Radford Hall is reconstructed Jacobean with wood carvings, completed by travelling Hugenots in 1622, rebuilt in Victorian times and now divided into several dwellings.

The Knowle lock flight, below which many powered craft turn.

The *Prince Regent II*, a luxury Edwardian-style cruising restaurant, is based below Radford Bottom Lock. Above this is Offchurch Viaduct, which carried the former Leamington to Rugby railway. More locks follow. There are shallows on the left before Fosse Middle Lock, above which is the B4455, the Fosse Way Roman road.

Views over the valley of the River Itchen become extensive, first to the north and then also to the south as the Bascote Flight of locks is

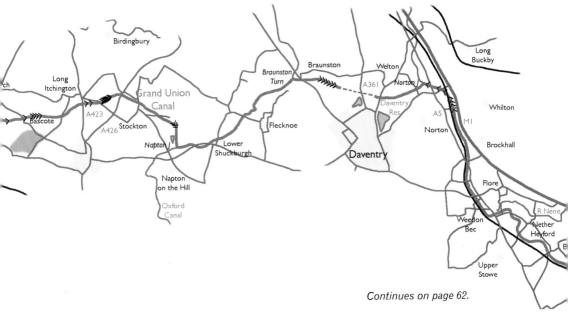

Continues on page 62.

Above and below: Kingswood Junction and the spur to Stratford-upon-Avon.

Continues on page 64.

Shrewley Tunnel with its unique horse tunnel climbing away.

climbed, a flight including the only staircase pair on the Grand Union Canal main line.

The spire of the 13th century church in Long Itchington was struck by lightning and collapsed during a service in a 1762 storm. The village has many 17–18th century half-timbered houses and an impressive stand of poplars around the pond. Wulfstan was born here in 1012, going on to become Archbishop of Worcester and St Wulfstan. Elizabeth I stayed at the Tudor House with the Earl of Leicester. Quarries produce a good supply of large fossils, of which the most notable in 1898 was a 5.8m-long near-perfect apatosaurus, 9m down and 20 million years old.

A reservoir has become a stocked angling lake beside the Kaye's Arm, dug to serve the cement works. Before Stockton Bottom Lock is the popular Blue Lias public house with its apatosaurus sign and flocks of marauding Canada geese, used to being fed by customers.

Next to a marina and Boat Inn, the A426

crosses Birdingbury bridge, dangerous because it is a fast road with a hump hiding the public house car park exit and the minor road junction on the opposite side. This is one of a couple of places on this road where useful high intensity signs have been installed, lighting up to warn fast traffic, although their effectiveness may be undermined by the fact that they are so commomplace.

Originally, it was planned to take a direct line from Warwick to Braunston but was later decided that money could be saved by using what was already there. From Napton Junction or Wigram's Turn, with its recent marina, the Grand Union Canal uses a section of the Oxford Canal. Traffic heading for London by the quickest route now travels along in the opposite direction from when it was first dug. It is a wide contour canal with excellent views both southwards to a ridge of high land and northwards over the valley of the returning River Leam.

Beyond Flecknoe, the former Leamington to Daventry railway crossed, to be followed a little

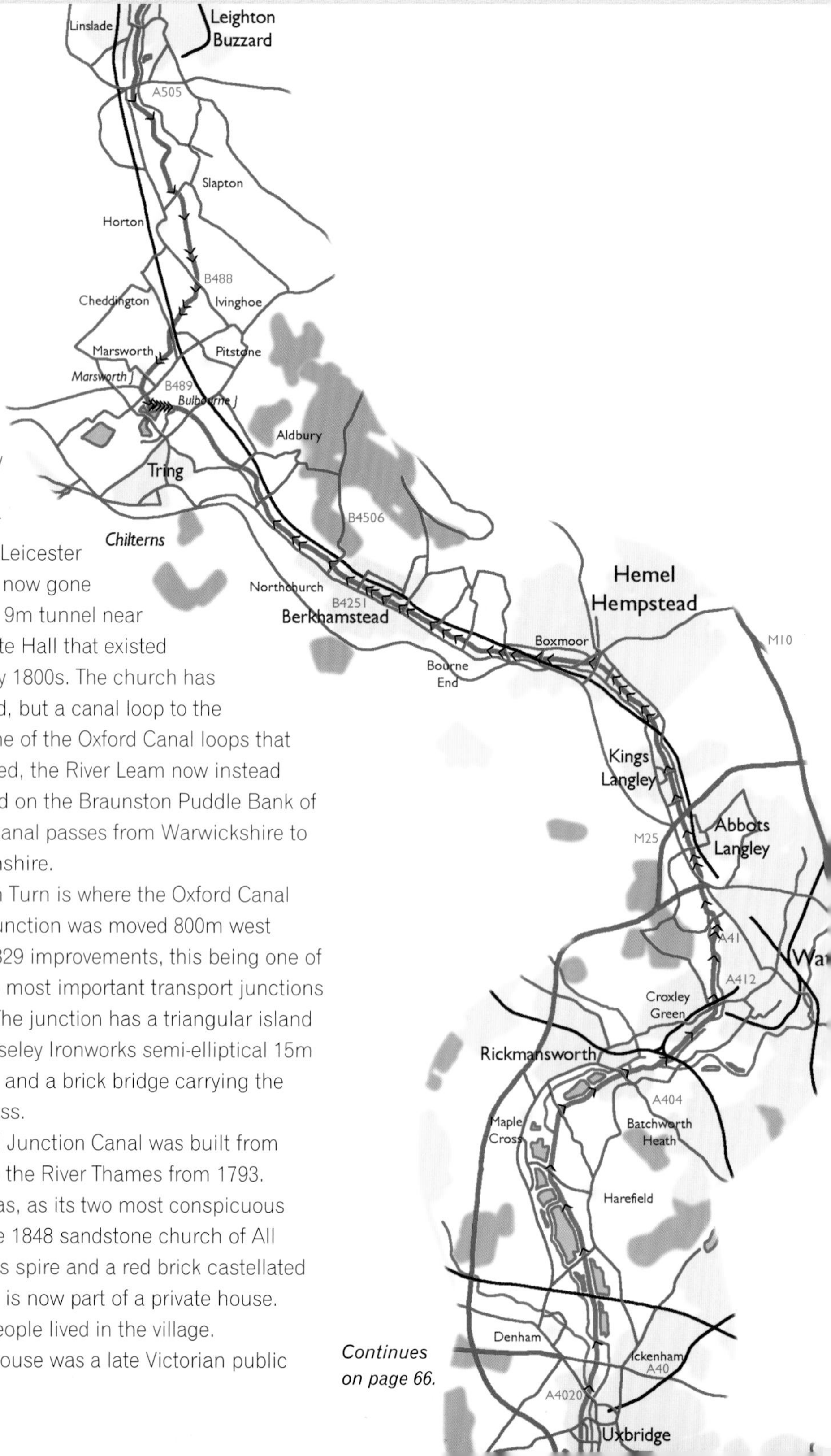

while later by
the crossing
of the former
Aylesbury to Leicester
railway. Also now gone
is the former 9m tunnel near
Wolfhampcote Hall that existed
until the early 1800s. The church has
been restored, but a canal loop to the
south was one of the Oxford Canal loops that
was eliminated, the River Leam now instead
being crossed on the Braunston Puddle Bank of
1830 as the canal passes from Warwickshire to
Northamptonshire.

Braunston Turn is where the Oxford Canal
leaves. The junction was moved 800m west
during the 1829 improvements, this being one of
the country's most important transport junctions
at the time. The junction has a triangular island
with two Horseley Ironworks semi-elliptical 15m
span bridges and a brick bridge carrying the
towpath across.

The Grand Junction Canal was built from
Braunston to the River Thames from 1793.
The village has, as its two most conspicuous
buildings, the 1848 sandstone church of All
Saints with its spire and a red brick castellated
windmill that is now part of a private house.
Many boat people lived in the village.

The Mill House was a late Victorian public

Continues
on page 66.

house, now extended and featuring canal paintings, photographs and prints.

A cast-iron bridge crosses what was the Oxford Canal until the line was shortened. A heritage centre has been housed in the 1796 Stop House, which charged excessive tolls to use the Oxford Canal. British Waterways' decision to depart from the historic building proved extremely controversial with canal users.

The Grand Union Canal built reservoirs for backpumping up the lock flight, rather than losing water to the Oxford Canal. The 1897 pumphouse replaced an earlier beam engine, sited in a field on the right. The reservoirs have become one of the largest marinas in the country, with buildings to house blacksmiths, harness makers, ropemakers and carpenters. It was the site of the former Braunston Boat Show, the main inland waterways festival.

Six locks in the flight take the canal up to Braunston Tunnel Gift Shop and the 110m summit level. On the way is the Admiral Nelson

pub, which is home to Northamptonshire skittles. It was used to film the *Inspector Morse* episode *The Wench is Dead* but has its own story in the form of a black figure who walks through a blocked-up doorway, causing glasses to rattle on the bar and pictures to fall down.

The summit pound brings moorhens and, more importantly, the 1.9km Braunston Tunnel. It was cut through an ironstone outcrop although 300m of it had quicksand. Setting-out errors resulted in an S-bend in the middle so that it is not possible to see through it. This added to the friction in the endless wire loop installed in 1870 and driven by a steam engine at the Welton end. This was short-lived. From 1871 to 1934 the tunnel was operated by steam tugs that started at hourly intervals from alternate ends. One of the leggers superseded by steam tugs was 75 years old and had worked in the tunnel for 44 years.

The walking route is unmarked. At first it is obvious enough, a track climbing up past rape fields and passing over the English watershed

The Hatton lock flight, the best locks on the Grand Union Canal.

before dropping down near Drayton Reservoir to reach the A361 via a kissing gate that is badly overgrown with stinging nettles. Opposite is a recent concrete farm track. When the track reaches a grass-filled cattle grid and rises to the left it is necessary to take an unmarked footpath to the right.

In 1793, permission was obtained to build a canal link to Daventry. There are now plans for the work to actually go ahead.

At Norton Junction the Leicester Line of the Grand Union Canal leaves, observed by a single-storey 1914 toll house and by two Victorian cottages, which were occupied by the Salvation Army who helped the boat people.

From here the canal gets very busy for a while as successive major transport links are met. The first in time and position is the A5, the Roman Watling Street, immediately below Buckby Top Lock where, until 1978, a shop used to sell decorated Buckby water cans. It still has the New Inn, the only public house remaining, where there were formerly seven for the six locks. It also has a choice between the two worst portages on the canal. When not obstructed by the lock balance beams there is a very narrow flight of steps leading down between vertical walls right next to the A5. These are probably not wide enough to allow a person to stand beside a canoe and there is no platform at the bottom from which to launch. The much more recent People Pipe under the A5 no longer floods as it did when it was first installed, but the steps down are at right angles to the rest of the subway so it is necessary to lower a canoe in at a steep angle over the handrails at the end, hoping nobody else is using it. Once beyond the A5, launching is not a problem.

Next to cross the canal is the West Coast Main Line, here the world's busiest main railway, which follows the canal closely to Blisworth and does

Above and below: Contrasting widths near Lower Shuckburgh.

not finally leave until Abbots Langley. Guardrails were fitted to stop horses falling in the canal when they were frightened by steam engines.

A location as far as possible from the coast but with a good new transport system was selected here for the 600m Weedon Military Dock. It had a yellow brick gatehouse with a portcullis; 12 magazine gunpowder stores for 5,000 barrels separated by blast buildings filled with earth; much small arms ammunition; 800,000 weapons (including 250,000 muskets and 30 pieces of field artillery); accommodation for two regiments; barracks for 500 soldiers; stabling for 200 horses; a riding school; and, later, a hospital. There were three large pavilions for the king and cabinet, including a royal pavilion for George III in case of a Napoleonic invasion. Although it was not used after 1920, it was believed to have been a planned staging post for Princesses Elizabeth and Margaret in the event of a Second World War invasion, from where they would have been taken to Prestwick and flown to Canada. When the railway line was electrified in 1965 the army pulled out and the entry arm was partly filled in and built upon. It is now private property, being used for industrial storage, although it has Grade II buildings. In 1991, a boatyard at the end put up a notice saying that they had been there 25 years and warning people not to move in opposite and then complain about the noise.

As the canal turns away from the open rolling Stowehill, the railway carries on and blasts straight through it; canal-style brick vent shafts are visible on the

Braunston Bottom Lock and covered dry dock.

Rural views at Bugbrooke.

hillside above the tunnel. The canal bores under the A5 again and passes the Narrow Boat and a small marina to arrive at a bridge near Flore.

From Flore the canal continues through completely rural countryside as it moves away from the River Nene, the A5 and the M1, disturbed only by the West Coast Main Line at intervals.

At Nether Heyford the church has a monument to judge Francis Morgan who pronounced the death sentence on Lady Jane Grey and then, filled with remorse, committed suicide in 1558. The village has a 2ha tree-lined green and a large Roman building to its east.

Bugbrooke has fine 18th century houses, a Baptist church of 1808, a 12th century chapel in parkland and a 14th century parish church with a 15th century wooden screen and a poem in the bell-tower warning against improperly dressed bellringers. The village had the first soap factory in England, as well as a brickyard, bakeries and mills. From the Wharf pub the canal

crosses the Bugbrooke valley aqueduct.

These Northamptonshire villages have an air of quiet affluence. Gayton is another with fine 16–19th century stone houses and a large church with an ornamented tower as well as an old manor house site.

At Gayton Junction the Northampton Arm leaves to drop down to the River Nene, currently the most practical route for many powered craft to reach the rivers of the east of England. There is a Canal & River Trust depot at the junction and a roving bridge across it. The A43 crosses the canal, as does the railway which then crosses the old A43 line north of Blisworth on a Robert Stephenson bridge where particularly good attention to detail is much in evidence. A tank in a garden previously supplied canal water to railway water troughs.

Blisworth has thatched houses with brown Blisworth stone and light Northamptonshire sandstone in alternate layers for decorative effect, a benefit derived from once having had tramways bringing stone from quarries to the canal. It begins at Candle Bridge, where boatmen bought their tallow dips for Blisworth Tunnel. St John the Baptist church dates from the 13th century and stands behind ornate cast-iron graveyard gates. The Grade II 18th century former Sun, Moon & Stars Inn backs on to the canal and Blisworth also had a mill.

The village is left behind but it has one more superlative feature to offer: Blisworth Tunnel. At 2.8km long, it is the third-longest canal tunnel in Britain, being surpassed only by Standedge and Dudley. Opened in 1805, it posed the greatest technical difficulty on the whole canal. The first attempt to cut through the ironstone outcrop in 1793–1796 had to be abandoned because of floodwater and a collapse that killed 14 men. The current line is a little to the west, although a ghostly candle-lit junction has been seen at the

The great warehouse in Blisworth.

accident side by some boaters. In 1977, part of the invert was found to have lifted by as much as 1.2m so the central section was lined with concrete segments over 1982–1984, during a five-year closure, yet it still showers water and the walls are covered with calcite.

Although the walking route is not marked, it is followed by a road for most of the way, although visibility is not good and there is no footway. At intervals there are tall brick vent columns, the tunnel having been built originally from 19 shafts lined up on Stoke Bruerne church tower, plus the end portals. These construction shafts were reopened after two boatmen suffocated in 1861 following a collision and five children had to be revived after a trip through in 1896, after which women and children were required to walk over. Towards the southern end the road bends right and a track ahead descends to the canal. On the way it passes through an embankment that once carried a tramway from Tiffield to Roade. Before the tunnel was built there was a tramway

connecting the two canal ends, the first iron railway in the south of England.

At the southern portal, one of the concrete lining rings has been set into the sloping bank to show the size to the hordes of spectators visible on any summer weekend, this being the start of Stoke Bruerne. Split by the canal, this is probably the country's best example of a canal village. Built in brown Blisworth limestone, the village had a boat children's school; Sister Mary Ward had a boat people's surgery for 30 years until the mid 1950s; and it was the site of what was probably England's last handmade ropeworks.

Stoke Bruerne Waterways Museum is located in a restored 1840s stone cornmill, used until 1913. Opened in 1963, it was Britain's first canal museum, displaying boats, engines, costumes, documents, photographs, paintings, models, a cabin, cabinware, china, utensils, brasses and signs. The museum shop is in an old beam engine house. The Old Chapel Gallery and Crafts Gallery is a museum of rural life and shows the work of up to 60 artists using wood, jewellery, ceramics and textiles. Rookery Open Farm, set up for children, is close by. There are tea rooms in old boat horse stables. The Boat Inn has been run by the same family since 1877 and is one of the few with Northamptonshire skittles. Over the village stands the Norman tower of the Perpendicular church. The village was used for filming some of the 1997 television series, *True Tilda*.

One of the Blisworth tunnel vent shafts.

The Navigation is a newer public house but equally convenient for the canal.

Stoke Park House was built in 1629–1636 as the first Palladian house in England. Designed by Inigo Jones, it had flanking pavilions, a colonnade and fine gardens, but was severely damaged by fire in 1886.

The Stoke Bruerne lock flight is crossed by the A508, beneath which it is like an art gallery. Framed mosaics by local schoolchildren on canal and other themes are mounted on both abutments of the bridge, a striking discovery for an unsuspecting passer-by.

The canal crosses two arms of the River Tove on the level and then follows its valley down to the River Great Ouse. The river passes over a long weir, above which the towpath is supported on brick arches. This design feature is seen again elsewhere but not as extensively as here.

On a hill overlooking the canal, Grafton Regis has a 13–14th century church with a Norman font, fine rood screen and many monuments to the Fitzroy family. Regis was added to the name following the visit of Henry VIII and Anne Boleyn to the manor house, home of Elizabeth Woodville, who had married Edward IV here in 1464 and become the mother of the Princes in the Tower. It was destroyed in the Civil War and is now a hospital. Charles II made his illegitimate son, Henry, the Duke of Grafton. The White Hart also has Northamptonshire skittles.

There are extensive views over the Tove valley. In particular, on the skyline 4km away is the Norman church of St James in Hanslope, which boasts the finest spire in Northamptonshire.

A sign warning visitors of elderly ducks crossing is the introduction to Yardley Gobion although it could equally be making reference to herons. The church dates from 1864 but there are older buildings of note. A 14–15th century pottery has become a public house, confusingly called the Coffee Pot.

North-east of the Navigation Inn at Thrupp Wharf is Castlethorpe, its motte and bailey to be found behind the Norman church of Sts Simon & Jude. The church tower collapsed in 1700, to be

The long arched weir feeding the River Tove.

replaced with a Georgian one. Again, there are fine views over the Tove valley.

At Cosgrove Priory the story goes that the daughter of the house fell in love with a shepherd. Her parents did not approve and had him deported on a false charge of sheep stealing. The daughter drowned herself in the millrace but her ghost may still be seen at the full moon.

Entry to Cosgrove, another village split by the canal, is under Solomon's Ornamental Bridge. Dating from 1800, in Gothic style, the bridge is the most elaborate on the canal. St Vincent's Well has a high iron content, good for eye problems, and is protected by Act of Parliament. The canal passes the Victorian rebuilt church with its 14th century weathercock and loops round the Barley Mow and the bay-fronted 18th century Cosgrove Hall, in front of which a Roman bath house has been excavated.

The canal runs straight across the valley of the River Great Ouse on a 1.4km long embankment, originally crossing the river on a stone aqueduct of 1805, although this collapsed three years later. While reconstruction work took place a flight of locks was built down to the river and up again on the west side although this was problematical

Beware, Yardley Gobion ahead.

Bill Bryson said 'I didn't hate Milton Keynes immediately.' It took a couple of hours of walking, trying to find the town centre with its 720m long mall, before he decided it was even worse than Gateshead's MetroCentre. Other views are divided. Clearly, many people like the little boxes hidden away in tree-screened cul-de-sacs and few places of this size have such easy rush-hour driving. The town is built in a loop of the canal, which is largely parkland where it does enter the built-up area.

At the start of New Bradwell, a Victorian railway workers' town, is the Grafton Street aqueduct of 1991, the first new aqueduct in England for 50 years and one of the most complicated pieces of civil engineering in Milton Keynes. Near it is an 1803 sandstone smock windmill, sited by the canal authorities for transport purposes, used until 1871 and now restored. An 1858 church by Street contains Victorian stained glass but a Norman chancel arch comes from a ruined church in Stantonbury. There are also the remains of Bradwell Abbey, founded in 1154 by the Benedictines.

After Newport Pagnell Road bridge, its abutments spray painted with bright butterfly murals, the canal is back in farmland.

The canal turns sharply under a former railway bridge and passes the 14th century St Andrew's church in yellow stone with a 12th century tower, Georgian box pews and fine 19th century stained glass. The late 17th century limestone Linford Manor, with Greek temple, almshouses with Flemish gables, two-storey school section and farm buildings, has become the Courtyard arts centre with pottery, jewellery and silversmithing.

The Newport Pagnell Canal served marble quarries from 1817 to 1864, became the Newport Nobby Railway until 1967 and is now the Railway Walk with nature reserve. An old bridge was retained after residents protested against its demolition. The old Wharf Inn at Linford Wharf has become a private house.

when the river was in flood. The remains of the locks and the old aqueduct remain in the bushes. In 1811 they were superseded by the Iron Trunk aqueduct, which has less than 200mm of upstand on the west side and which locals called the Tank or Pig Trough. A scheduled ancient monument, it was one of the first of its kind in the world. Beneath the canal there is a narrow oval pedestrian tunnel. A 1981 plan was to place locks on the Great Ouse to make it navigable to powered craft to Bedford. If the Bedford & Milton Keynes Waterway fails, the plan may be revived. Crossing the aqueduct takes the canal from Northamptonshire to Buckinghamshire; the canal then follows the river to Haversham.

The embankment ends at some houses, one of which has a dovecote, and the Galleon public house. Behind this are an 1815 church in Norman style with a 1729 rectory, the remains of a Norman motte and bailey and the site of the medieval Wolverton Village.

This is the start of Milton Keynes, Britain's largest new town, built on 89km^2 of farmland from 1967 and featuring millions of newly planted trees and sweeping American-style roads.

By Gulliver's Land and Willen Lake in its 1.1km² park are the first Buddhist peace pagoda in the West and St Mary Magdalen, the only Wren church outside London. This reach is the starting point for the proposed Bedford & Milton Keynes Waterway. At 32km long, a broad canal, it will connect the canal network with the East Anglian rivers and will be the longest new canal in Britain for over a century. However, previous attempts in 1811 and 1892 failed and success is not assured this time.

One of the surprising things about Milton Keynes is its name. Rather than being called after Bletchley, Wolverton or one of the towns it incorporated, Milton Keynes uses the name of a village that manages to retain clear space between itself and the new town, standing on the far side of the River Ouzel or Lovat, the valley of which the canal is now to ascend.

The Old Swan Inn at Woughton on the Green was used by Dick Turpin and his ghost has often been seen on his horse, Black Bess. One night, while being pursued, he apparently quickly reshod his horse with the shoes on backwards to fool his pursuers. Assuming he had the tools and noise was not an issue, how long does it take to reshoe a horse? Would such an experimental procedure be the best use of that time unless the pursuers were so far away that a more leisurely departure could have been made?

Woughton Park and Walton Hall have been home of the Open University since it was set up in the 1960s to provide correspondence courses for those not able to attend full-time study at university.

The George Amey Centre for outdoor education operates from a former lengthsman's house. The Plough serves those in need of liquid refreshment.

Before Fenny Stratford there is a house with a garden seat which uses a cartwheel as its back, attractive to look at but probably less comfortable to use, especially with its large boss in the centre. The lock is also awkward to use as it has a swing bridge across the centre of the chamber. On the other hand, the rise is only 300mm, the smallest on the canal. The bottom pound leaked but only near the surface of the canal. Rather than solve the leakage it was decided to install an extra lock and accept a slightly lower level in the bottom pound. The old Grade II northern engine pumphouse was originally steam powered and has a craft centre, gift shop and tea room, run by handicapped youths, or the Red Lion is close by.

The Bletchley to Bedford railway crosses and the following bridge is on the line of the Roman Watling Street, accompanied by the Bridge Inn. On November 11th six cannons will be heard, the Fenny Poppers, celebrating the dedication of the red brick St Martin's church. It was built in 1724–1730 in early Gothic revival style. The tradition was begun in 1760 by antiquarian Dr Browne Willis, the lord of the manor.

A golf course and deer farm are passed to reach Soulbury Three Locks or the Stoke Hammond Three. The Grand Union at the Three Locks public house is located in the former stables. The area is haunted by a woman and child, the noises of a lock are heard operating at night, pram wheels squeak and a woman calls for her drowned family.

There are plenty of very fine views over the River Ouzel valley near the border between Bedfordshire and Buckinghamshire, interspersed with gorse as the canal begins to follow the river's exaggerated meanders.

Leighton Buzzard is a 17th–19th century town of half-timbered houses including the 17th century Holly Lodge. It has a Georgian high street plus a pentagonal market cross of 1400. All Saints' church of 1288 has a 58m central spire, 15th century wooden angel roof, sanctus bell, 13th century font, misericords, brasses, medieval lectern and fine stained glass by Kempe and others. Medieval graffiti is to be seen throughout the church, the most famous of which is the 'Simnel' scratching on a pillar. This is thought to depict a scene from the local story of Simon and Nellie, who created the first Simnel cake. The bounds are beaten on Rogation Day. The town name may be from the Old English *leac-tun*, a vegetable farm, belonging to the Buzzard family.

The Grove Lock cottage has been converted to a public house. At the following Church Lock, the 14th century chapel with bell turret has been converted to a private house. There is a large flooded sand pit to the left. The Chilterns are becoming more prominent ahead. Bury Farm has an equestrian centre.

The locks on the ascent of the Chilterns are mostly in open country. Slapton Lock or Neale's is not too far from Slapton, where the Perpendicular church has notable 15–16th century brasses.

Horton Lock gives the best view of the scarp slope of the Chilterns before they become lost to the foreground scenery. To the east hot air balloons and gliders can be seen flying from Dunstable. The 150m Whipsnade white lion was cut in 1935 in the chalk of the Dunstable Downs to promote the zoo, and is the largest chalk carving in the country.

As the climb continues the locks get closer together and the route twists and turns more frequently. The two Ivinghoe Locks are followed by three Seabrook Locks, the middle of which has the best preserved of the back-pumping northern engine houses ('northern' meaning those north of the Chilterns). To the east are Gallows Hill, with an array of aerials, and Ivinghoe Beacon, the highest point on the Dunstable Downs and eastern England at 230m. The beacon was prepared for Elizabeth I to summon men in the event

Solomon's Ornamental Bridge at the entrance to Cosgrove.

An avenue of poplars at Water Eaton.

of a Spanish invasion. Ivinghoe probably gave Sir Walter Scott the title of his *Ivanhoe*. Beyond Pitstone is Britain's oldest post windmill, the Pitstone mill of 1627. At one time it was owned by the canal company but they sold it in 1842 to the Ashridge estate. It was badly damaged in a squall in 1903, was restored in 1962 and now belongs to the National Trust.

Marsworth, also known as Maffers, is one of the most attractive points on the canal, especially where the B489 crosses by the closed White Lion at Startop's End. Bluebells Tearooms are in a lock building. The adjacent car park is notorious for theft. Tring Reservoirs were built in 1802–1839, with water pumped up to the 120m summit level where the channel is deeper than usual for storage.

Moving from Buckinghamshire into Hertfordshire the canal is now in Tring Cutting, 2.4km long and up to 9m deep, the site of a wartime bomb strike.

Beyond Aldbury, with its village green, stocks and pond, is a 33m Doric column and urn monument of 1832 to the 3rd Earl of Bridgewater, the canal pioneer who owned the 16km² Ashridge estate on which it is sited. The estate boasts good oak and beech woodland and has been the location of many films.

Cow Roast is a corruption of cow rest, where there were pens for the use of drovers bringing cattle from the Midlands to London. Earlier roasting took the form of iron working in a Celto-Roman site. The lock begins the continuous descent to the River Thames. Crossing the bridge is the Icknield Way Path. While there are various routes suggested for the Icknield Way, the Ridgeway and the current footpaths that take their names, it is clear that the route generally ran along the Chilterns in this vicinity. This Neolithic route was founded around 4000–2000 BC, the oldest road in Europe.

The proliferation of lock names continues. The next one, below which is a magnificent copper beech, is called Bushes, Old Ned's, Awkward Billy or Crooked Billet. The two Northchurch Locks, by a park and playground, are also called the Gas Two from a former gasworks. Berkhamsted Top Lock is also called Broadwater and is positioned above a striking bridge painted black, red and gold.

The market town of Berkhamsted drew its wealth from wool in the Middle Ages but also became the port of Berkhamsted. It has four 17–18th century coaching inns and Berkhamsted School, founded in 1541. Graham Greene was born here and went to the school, where his father was headmaster, as a result of which he was bullied and was sent away for six months. Sir James Barrie visited friends with five sons in Berkhamsted, who inspired *Peter Pan,* in 1904.

Berkhamsted Castle has a double moat, a tall 11th century motte and bailey, curtain walls with D-plan towers and a square tower-like keep. It was built by Robert of Mortain, the half brother of William the Conqueror, who received the oaths of allegiance of the Saxon kings here in 1066. In the 12th century the timber was replaced with stone and flint walls, breached by the stone-throwing machines of Louis of France and the barons opposed to King John in 1216, who took the castle after a fortnight's siege. Projecting platforms of the outer earthworks may be siegework from this campaign. Brick buildings were added later.

Thomas à Becket spent time here as Chancellor. The Black Prince lived here from 1336 and it was a favourite royal residence until Elizabeth I. It was visited by Geoffrey Chaucer and Henry Tudor and was used to imprison King John of France but had fallen into ruin by 1495, the materials being used in town house building. It still belongs to the Prince of Wales.

The Crystal Palace pub may have obtained its name from locally-employed William Paxton, the brother of Joseph Paxton who designed its larger namesake. Opposite is a totem pole carved in 1967 by a Kwakiutl Indian from Vancouver that belonged to a Canadian timber company. However, when they moved to new premises they were not allowed to take it with them. A former boatbuilding yard launched craft sideways into the canal here. Berkhamsted Bottom Lock, by the Boat, and the following lock, by the Rising Sun, were both referred to as Sweep's because chimney sweep Eli Oliffe kept a boatman's store. Dalton Wharf, which follows, had the first factory for producing sheep dip.

The 13th century St Peter's church, one of the largest in Hertfordshire, was restored in 1871 by Butterfield and has several brasses and an east window dedicated to William Cowper, whose father was rector when he was born in 1731. The 16th century Incent House has been restored but the 17th century Sayer almshouses are intact because of a bequest by Charles II's chief cook.

Hemel Hempstead Cricket Club's ground is screened by poplars as the canal approaches Boxmoor Bottom Lock. The A414 crosses Two Waters Road Bridge, adjacent to the confluence with the River Gade, which is now followed.

The West Coast Main Line crosses for the last time and the canal wanders down to Kings Langley. This is the site of a former papermill and the Ovaltine works and now features a 220kW wind turbine.

When the foundations were being dug for the Rudolph Steiner school in 1970 they unearthed the remains of a 13th century wine cellar. There was a 13th century friary here and Henry III had a palace. Edmund de Langley, the first Duke of York and brother of the Black Prince, was born at the palace in 1341 and is buried in the Norman All Saints church, as is archaeologist Sir John Evans. The Saracen's Head is rather more recent, dating from the 17th century.

A royal hunting lodge site and a moat are overshadowed by a 440m viaduct carrying the M25 over the canal, river, road and railway. The London Orbital is the world's longest city bypass and Britain's busiest motorway. Junction 20 is on the west side of the canal. For practical purposes many people now regard the M25 as London's boundary. Canal piling by the viaduct is tied together with a mixture of standard- and narrow-gauge railway line.

Home Park Farm Lock is on the outskirts of Abbots Langley. In 1931 the Ovaltine Egg Farm was established here, based around a thatched and half-timbered building. It is now a housing estate. Nicholas Breakspear was born in the town, becoming Pope Adrian IV in 1154, the only English Pope.

Watford, a market town in the Middle Ages, is the largest town in Hertfordshire.

To the west of Lady Capel's Lock is Grove Park, formerly 77ha of gardens for the Earl of Essex in the 17th century, landscaped by Stubbs. It is now a golf course with chestnuts, oaks, ashes and a country park feel. The balustraded Grove ornamental bridge was a demand made by the Earl of Essex when the canal was built.

Grove Mill was formerly the home of the Earls of Clarendon, and is now exclusive flats with some interesting pieces of machinery placed like sculptures on the lawn by the canal. The canal has some tight bends and a wooden signal box was built to control canal traffic, the towpath diverted to the east side. The local fauna might include llamas.

The Globe Inn on the northern edge of Linslade.

The lower of the Cassiobury Park Locks has a Victorian cottage that was wrecked by vandals. British Waterways were in the process of demolishing it in 1970 when a buyer arrived, using an Alsatian to stop the demolition and later going on to restore it. A notice on the fence warns about the danger of crocodiles.

Ironbridge or Watford Lock has seats made from balance beams that look more substantial than the metal ones on the lock gates. The lock is haunted by a man who opens the bottom gate, whatever the balance beam design.

Watford has an annual carnival at Whitsun, perhaps best reached beyond Cassiobridge Lock over the bridge carrying the Metropolitan Line, the canal's first meeting with the London Underground system.

An industrial estate occupies the site of Croxley Mill, built by paper manufacturer John Dickson in 1830 and demolished in 1982. This used coal brought via the Ashby-de-la-Zouch Canal and esparto grass also delivered by canal. Dickinsons were probably the largest customer for Grand Union Canal transport.

At the southern end of Croxley Green, Croxley Hall Farm's medieval tithe barn is the second-largest in the country. The Metropolitan Line crosses back, without the planned depot that was refused by the local authority. It is followed by Lot Mead Lock, also known as Walker's, Beasley's or Cherry's after various overseers.

A 15th century house, built for the Archbishop of York, became Cardinal Wolsey's country seat, reconstructed in Baroque style in 1727 by Leoni as the most splendid 18th century mansion in Hertfordshire, with marvellous interior frescoes.

The grounds of the house were landscaped in 1758 by Capability Brown.

Rickmansworth was a market town that got its charter from Henry VIII. When Elizabeth I died it was local landowner Robert Carey who travelled to tell James VI of Scotland that he was also king of England. After getting married, William Penn lived here from 1672 for five years before founding Pennsylvania. There is a brick and timber priory, the half-timbered vicarage is partly medieval and St Mary's church was rebuilt in the 19th century in a way that retained its early Gothic windows. Frogmoor Wharf was used to build and repair boats – many of them canal boats but also some targets for Portsmouth's Whale Island gunnery school, a total of 212 large boats in addition to punts and pontoons. Training of mini submarine crews was undertaken in the canal at Rickmansworth. Batchworth Lock is at the confluence with the Rickmansworth Branch, the dredged part of the River Chess, which served the town wharves, gasworks and gravel workings. It has a rare wide beam lift bridge. In 1973 there was a study for a barge depot to serve road traffic from the M1 after lower canal enlargement.

A restored lock cottage with canal artefacts and an ivy-clad Georgian house are to be found at Stocker's Lock. Otherwise there are 16th century farm buildings with steep grazing to the south and some very smart horses.

The canal now moves from Hertfordshire to London. Chalk quarries are cut into the hill and orchids grow in them. Springwell Lock has been used for filming *Blake's Seven* and *Doctor Who* episodes. Springwell Lake is also used for angling and birdwatching. It has a nature trail with the Springwell Reedbed Nature Reserve at Maple Cross. Its industrial estate is by the canal and Lynsters Lake. Lilies appear in the canal.

Black Jack's Lock takes its name from a black man employed by a mill owner to harass boat people and stop night-time toll evasion. He

One of the Seabrook locks.

acquired a stock of windlasses that he stored in a hollow tree. His efforts were not universally appreciated and he was murdered by the boaters, his ghost still haunting the area.

Beyond Widewater Lock at South Harefield is the Horse & Barge public house and a pillbox. Harefield Marina is in a former gravel pit where over 50 working boats were sunk in 1958. Hoveringham Lake is used for a watersports centre and carp angling.

The west side of the canal has a golf course, Denham Country Park, a picnic area, a visitor centre, Tudor half-timbered cottages and a church with a 500-year-old Judgement Day mural. An aqueduct takes the canal over the Frays River, a diversion of the River Colne at Ickenham. Below is Denham Lock, the deepest on the canal at 3.4m, used for filming a *Mogul* episode. Denham studios opened in 1936, producing such films as *Brief Encounter, Robin Hood* and, in 1963, the most famous canal film of them all, *The Bargee*.

The Misbourne river joins and there is a recent gravel wharf with a conveyor for loading barges. The A40 crosses Oxford Road Bridge, immediately before becoming the M40. Beyond the bridge a gas main crosses over. Gradually the canal becomes more industrial. Sanderson Fabrics have their playing fields at Willowbank. Uxbridge Lock has a roving bridge. A flourmill that was once run by Quakers has been redeveloped as flats.

Uxbridge is a mixture of half-timbered, Georgian and modern buildings, less placid than it appears. In 1555 three people were burned at the stake here during religious unrest. In 1630 there were riots against tolls imposed by the Countess of Derby, the lady of the manor. It was from the RAF headquarters here that Air Marshal Lord Dowding directed the Battle of Britain. The Swan & Bottle is opposite the Crown & Treaty Inn of 1576, which was the home of Sir John Bennet, who hosted the unsuccessful Treaty of Uxbridge in an attempt to end the Civil War, this being a Roundhead base. The building's panelling went to the Empire State Building for a time but has since been returned.

Rank Xerox have tower blocks by the canal. At one time Fellows, Morton & Clayton had one of the largest working fleets on the Grand Union Canal with boats triple-moored here. Denham Yacht Station and Uxbridge Boating Centre have it easy today.

The General Eliott and a Royal Mail depot are by Dolphin Bridge while the B470 crosses above Cowley Lock. In 1896 a barge loaded with chalk drifted on to the cill, broke its back and sank while the crew were in the adjacent Malt Shovel, to which can now be added the Grade II Toll House tea rooms.

A 43km level pound runs from here to Camden. From 1801 there was a four-horse

Bulbourne Dry Dock.

packet boat service to Paddington, with precedence over all craft, operating from the Paddington Packet Boat.

Cowley Peachey Junction brings in the Slough Branch of 1882, one of Britain's newest canals, which is very straight but surprisingly narrow at the junction. The desire of canal enthusiasts to link this branch to the Thames or Jubilee River at Slough faces constraints in the form of buildings, the railway and the M4, so it must remain low on the priority list.

The canal turns east at Yiewsley as other traffic begins to intrude. The Reading to Paddington railway comes alongside at West Drayton station and Heathrow airport is 3km ahead, the main runways lying parallel with the canal, although a third runway might be just over a kilometre away. The UK's largest airport with the world's largest throughput of international traffic, it has the longest British civilian runway at 3.9km.

Parts of St Martin's church date from the 15th century, including the font. It also has fine monuments, among them the grave of Captain Billingsley, who was lost with the *Royal George* and 800 men in the Solent in 1782. Opposite an aerial is Stockley Park, a country park golf club and business park.

From the A437 the view is to the railway crossing, usually with a background of a procession of aircraft landing. A coffee smell comes from the 1929 Nestlé factory. The canal crosses over Yeading Brook aqueduct, the stream formerly being called Bulls Brook, and is then crossed by the A312 in a former brickfield area. Bull's Bridge Junction takes its name from the old name for the brook and is where the Paddington Arm leaves. Tesco have a supermarket on the site of the headquarters of the former Grand Union Canal Carrying Company depot, a place where a Bofors gun was manned during the Second World War and where there

Toll keeper's house at Bulbourne Junction, the start of the Wendover Feeder Branch.

Former British Waterways' workshops at Bulbourne.

was a school and social club for boat crews. The British Docks Board research centre for such concerns as tidal flows and erosion was located here with large models. Tickler's Dock was known as the Jam 'Ole and their coal was the last long-distance traffic on the canal, occasionally recreated with a Jam 'Ole Run.

The water collects increasing flotsam yet it is cleaner than might be expected, which is perhaps an indication of limited boat traffic. A typical dock has not so much houseboats, as

Apsley with its old warehouse and new footbridge.

floating houses. Opposite the Grand Junction Arms and next to a Porsche repair centre is a new building that complements a wharf crane in good condition.

The former Old Oak Tree is by the canal at North Hyde and the Lamb by the A3005 Wolf Bridge, named after a different public house. The Maypole Dock of 1913 served the Monstead margarine works, the world's largest, which is now owned by Quaker Oats. A gold dome to the north is a reminder that Southall has long had the largest southern-Asian population outside Asia. There is a significant number of coconuts in the water, evidence of Sikh funerals.

Beyond Norwood Locks the canal is crossed by the A4127 on Windmill Bridge, named as such because it was near the site of Southall Mill, famously painted in 1810 by JMW Turner. It was also called the Three Bridges, which is incorrectly named as it is a road bridge over a canal aqueduct over a freight railway, the complex designed by Brunel in 1859. Some of his broad-gauge track is now used as fencing although, unusually, there is an unguarded drop from the canal to the railway.

Grove Mill with old machinery decorating the lawn.

The Hanwell Flight of six locks, a scheduled ancient monument, was operated by tow tractors at one time and was used in the filming of the 1964 comedy film *The Bargee*. Alongside is St Bernard's hospital, which was once the county lunatic asylum. A conspicuously bricked-up arch was used for coal delivery by canal and rectangular holes were for fire hoses. The hospital produced all its fruit, vegetable and animal requirements and the surplus was taken away by canal. The flight has horse steps and tubular hitching posts.

At the bottom of the flight the canal joins the River Brent from the Welsh Harp feeder reservoir, deepened for barges for the rest of the way to the Thames, and the Brent River Park from the A40, crossed upstream by the railway over the 1837 Wharncliffe Viaduct.

Giant hogweed appears along the banks and a nature reserve has been built on a former refuse tip. Beyond Osterley Park's aviary, Osterley Lock is accompanied by new bird nest boxes at the first of three places where the canal cuts off bends of the river.

The M4 passes over and large houses begin to appear. On the right bank, beyond Wyke Green golf course, is Osterley Park House, a crumbling 1576 mansion built for Sir Robert Gresham, founder of the Royal Exchange. Robert Adam converted the property from 1763 to 1767, one of the most complete examples of his work. It has a richly decorated interior, Georgian furniture, wall paintings, sculptures, Gobelins and Beauvais tapestries and fine carpets. A 16th century stable is still in use. It is all set in a 1.4km² park and farmland with Pleasure Grounds, neoclassical garden, lakes, orangery and nature trails and was used for filming *The Queen, Burke & Hare, The Grass is Greener* and *Miss Potter*.

The Piccadilly Line crosses and the next bridge is Gallows Bridge, a fine cast-iron bridge of 1820, probably the first canal bridge by Horseley Ironworks. The inscription of 'Grand Junction Canal Co. 1820' was altered to 'Grand Union Canal Co. 1820' although the Grand Union Canal was not formed until 1929.

Beyond the M4, which is climbing up on to Chiswick Flyover, is Boston Manor House, a fine Tudor building of 1623 with 1670 Jacobean extensions, good plasterwork and one of the best English Renaissance ceilings.

The A4 Great West Road is next to cross, quickly followed by the Barnes Bridge to Hounslow railway. After these there is a Canal & River Trust depot with warehouse roof over the towpath and the canal, the last remaining overhanging warehouse roof of its kind. The rest of this reach has been dramatically rebuilt with a dazzling array of new apartments. At the far end of the reach with its Grade II toll house, one of London's smallest museums, is Brentford Gauging Lock, which is actually two locks side by side, beyond which the water is semi-tidal.

Below the lock is Brentford High Street bridge, carrying the A315, first built as the Roman road from London to Silchester. In the time of Edward I the bridge was kept in good repair by a toll levied on every Jew passing over. An Act of George IV set a penalty of penal servitude for life for wilfully damaging the bridge. Grounds Coffee & Wine Bar is one of the few conspicuous buildings that does not seem to be brand new, surprising as Brentford was the county town of Middlesex. JMW Turner was a local resident, as was Betty Higden in *Our Mutual Friend*. Lionel Wallace saw the door in HG Wells' *The Door in the Wall* when called to a House of Commons division while dining at Brentford.

The grounds of Syon House have seen plenty of activity over the centuries. In 1016 Edmund Ironsides defeated the Danes here. In the 1642 Battle of Brentford, Prince Rupert beat the Parliamentarians and the Civil War might have ended there but the Royalists blew up an ammunition barge and the noise was mistaken for gunfire. Syon House itself began as a monastery for the order of St Bridget and had the

dubious honour of being one of the first religious houses to be suppressed by Henry VIII. Perhaps it was just coincidence that dogs got into his coffin and savaged his corpse while it rested here overnight on the way to Windsor for burial in 1547, after it had been predicted that dogs would lick his blood for damaging the abbey. Katherine Howard was imprisoned here until she was executed, Lady Jane Grey was said to have been here when she was offered the crown, Charles I as a prisoner visited his children here and it was loaned to Pocahontas in 1616.

The building was an early Robert Adam restoration in 1762 of a 1547 house and has magnificent interior decoration, furniture and paintings. The 81ha of landscaping by Capability Brown includes an 1820 great conservatory by Charles Fowler with glass dome covering 280m^2 and free-flying birds inside, the London Butterfly

A wet evening at Cowley Lock.

Gallows Bridge was one of Horseley Ironworks' first for a canal. It was close to a gallows site.

House, a heritage motor museum, art centre, needlework centre, aquatic centre, one of the largest garden centres in the London area, a rose garden and 200 rare tree species. It is London's only ducal residence and has been the Duke of Northumberland's London base for over four centuries. The current duke's family were involved with Inland Waterways Association campaigning and running a working fleet on the canal in the 1950s. The grounds include a three-span, very early, wrought-iron bridge of 1790 over an ornamental pond, similar in profile to, but of different materials from, one built in 1812 by the Duke of Northumberland on his estate in Anwick. The venue was used for filming *The Madness of King George* and *Gosford Park*.

The Dock Road Bridge, a riveted girder construction by Brunel, carries Augustus Way but was built as a railway bridge for a large railway yard on both sides of Brentford Dock (some 30 sidings in all), which has gone. The dock, which had retaining walls held apart by horizontal arches, was open from 1859 to 1964 but has now been replaced by housing. This was the birthplace of the Bantam push tug.

Downstream there is a large fig tree on the left and, behind that, Fuller's Brewery Tap. There are high walls beyond with not even a ladder up in the vicinity of Thames Locks, not a very satisfactory situation as there is a weir off to the left. Brentford Gut is tidal and the locks can be used by pleasure craft for a two-hour period either side of high water or can be opened through at high water.

It was usual for a man to bring up a lighter on the tide, unpowered and with only a steering oar. One novice failed to make the turning and was carried on up the Thames, going down and up with the tide and finally getting it on the next pass, very much more tired for the experience.

The Brent Ford got its name from a sand bar produced across the Thames by its tributary and it was here that Julius Caesar crossed the Thames in 54 BC. Opposite is the World Heritage Site of Kew Gardens. The 1631 Kew Palace in red brick Dutch style was the smallest royal palace and was subject to flooding while George III spent his final years here in 1800–1818. Queen Victoria's parents married in the drawing room.

The Royal Botanic Gardens were begun in 1752 by Princess Augusta and cover 1.3km^2 with over 45,000 plant species in Victorian and modern conservatories, including the 1848 Palm House by Decimus Burton and the 1987 Princess of Wales Conservatory with its 10 climatic zones. The 1899 Temperate House is the world's largest surviving Victorian glass structure at 4,880m^2, 110m long and 19m high. It is also one of the world's most important, featuring the first use of wrought iron ribs in an architectural building. Kew's highlights include the world's tallest flagstaff at 69m, an orangery, follies, temples, statues, water features, the world's oldest orchid collection, lion trees from 1762, a Chilean wine palm that is the world's tallest indoor tree, herbaceous bedding, an arboretum, a treetop walkway 200m long and 18m high, an explorer train, rainforest to desert environments and a herbarium that is used as the international standard for plant identification, housing an eighth of the world's known species.

The Ovaltine narrowboat mooring at Brentford.

12 Grand Union Canal: Leicester Line

The Loughborough Navigation made the River Soar navigable up to Loughborough in 1778, the most profitable canal in Britain for a while. The Leicester Navigation continued this to Leicester in 1794. The Leicestershire & Northamptonshire Canal extended the line to Debdale Wharf in 1797 and to Market Harborough in 1809. It had originally been planned to go from Debdale Wharf to Northampton but technical

difficulties proved too great. Once what was to become the Grand Union Canal main line was built from London with a branch to Northampton, it was decided to change the plan and run from Foxton to Norton Junction instead, this Old Grand Union Canal being completed in 1814. It carried Nottinghamshire and Derbyshire coal, Mountsorrel and Quorn granite and through traffic but was damaged by coal-carrying on the Ashby-de-la-Zouch Canal and by the London & Birmingham Railway from 1838. In 1894, the Grand Junction Company bought the line from Leicester to London, expanding it to form the modern Grand Union Canal in 1926 and extending it to the River Trent in 1932. It is a contour canal and forms part of the Leicester Ring. It includes three of the dozen longest tunnels still in use and three of the longest half dozen lock staircases.

Four transport arteries from different eras squeeze together through the Watford Gap. The canal meets them in chronological order starting with the A5 Watling Street, first built by the Romans and improved as the London to Holyhead mail coach road. Next over is the West Coast Main Line on what is nearer to being a tunnel of concrete beams but that retains the facades of the original bridge with their decorative work picked out in colour, a welcome piece of preservation that is only really seen from the canal. Immediately on the right is the M1's Watford Gap Services of 1959, surprisingly unobtrusive but accessible for provisions or meals even without a vehicle. Britain's first service station, it had been intended originally for lorries only.

After a second marina and a field of caravans, seven locks (four in a staircase) raise the canal 16m to the summit level of 126m above sea level, the 20-Mile summit pound being one of the highest and most beautiful with some of

Thoroughly agreeable scenery at the foot of the Laughton Hills.

Distance
106km from Norton Junction to the River Trent
Highlights
Foxton locks, the largest example of staircase locks in Britain
The Mile Straight in Leicester
The National Space Science Centre
Museum of Science & Technology
Navigation Authority
Canal & River Trust
Canal Society
Foxton Inclined Plane Trust www.fipt.org.uk
OS 1:50,000 Sheets
129 Nottingham & Loughborough
140 Leicester
141 Kettering & Corby
152 Northampton & Milton Keynes

the most lonely canal country in Britain. It was planned as a wide canal to carry traffic between the Thames and the Trent but lack of water meant it had to be narrow over the summit, preventing a wide link between the south and north of England.

Noise from motorway traffic is blocked out to some extent by trees along both sides of the canal as the M1 passes over.

Parkland occupies much of the area over to Watford, where the 17th century Watford Court with Victorian additions was demolished in the 1970s. Still present is the 13th century church with interesting monuments.

The Northampton to Rugby railway crosses over as the canal arrives at Crick Tunnel. The 1.4km long tunnel is wet, having been built on a revised line to avoid quicksand, but there have been three roof collapses, and 26m had to be rebuilt in 1987. There is no towpath but there is a benevolent ghost that has been seen making tea in narrowboat galleys. Medieval ridge and furrow systems prevail over the southern end of

Flat Leicestershire fields to the north of Husbands Bosworth.

the tunnel. The correct route is back up a track to the right, along a minor road to Crick, first right down Horse Boat Lane, following the start of a bending footpath for West Haddon, and then cutting left across a field to a white sign above the towpath; a less-than-obvious route.

Crick, which would not otherwise be seen, has a manor with medieval vaulted cellars and a church with much decorative stonework. The rector of 1622–1624 was William Laud, later Archbishop of Canterbury.

Beyond Elkington the A14 crosses. The Jurassic Way footpath uses the towpath. Up near Elkington Lodge a dinosaur of a windpump stands on the hillside. The Hemplow Hills lead on to 169m high Downtown Hill, which is used for growing broad beans.

The fields to the left were used by Percy Pilcher, the first Englishman to fly. He was killed in 1899 when his hang-glider-like plane, *The Hawk*, crashed. A monument to him is in a field north of Stanford on Avon and Stanford Hall has a replica of the plane. Flights today take the form of gliding from the airfield between Welford and Husbands Bosworth and also the winter wildfowl on Stanford Reservoir. The site of Downtown

Village is on the left bank in front of the reservoir.

The canal crosses the River Avon and the Northamptonshire–Leicestershire border on an aqueduct. Views in the North Kilworth and Welford directions are extensive at what is close to the country's watershed. The River Avon flows westwards while the headwaters of the River Welland flow eastwards from just east of the canal. The River Avon is closely followed by the Welford Branch, which was built as a feeder from Welford, Sulby and Naseby reservoirs but has a lock and carried up coal and brought down lime. It was disused from 1938 until restored in 1969 and is a popular destination for narrowboats.

North Kilworth had the first wharf on the canal, receiving coal from the Warwickshire and Derbyshire coalfields. Anglo Welsh have a boat hire base here. The canal heads into a wooded cutting. St Andrew's 12th century church in North Kilworth is noted for its Jacobean Armada pulpit of Spanish oak.

Husbands Bosworth Tunnel is 1.1km long. This time the horse path is rather more obvious as it climbs over a hill, crosses the A5199 and winds over a footbridge across the former railway. Perhaps this is as well as there are fewer spoil heaps along the line of the tunnel for guidance. This tunnel also has a ghost, reports of which being very useful for keeping away strangers who might have sought work legging the working boats through.

The canal emerges into a cutting below Husbands Bosworth. All Saints church has a 14th century spire and there is a Georgian Bosworth Hall, partly from the 16th century, which hosts Husbands Bosworth Festival with traction engines and much more.

Foxton is one of the most fascinating points on the canal system. It starts where a tall lock keeper's cottage overlooks the top of the lock flight. There are two staircases of five locks with a passing place in the middle, the largest example of staircase locks in Britain. Built 1812, they drop the canal 23m from the summit level. Although they had side ponds to save water, they caused a tremendous bottleneck for traffic, so an inclined plane was added. Based on the one at Blackhill on the Monkland Canal, it had tracks at 1:4 and could carry wide-beam traffic, unlike the lock flight. It had two tanks, each 24m x 4.6m and 1.5m deep, weighing 220t apiece, balanced by wires passing round a single steam-powered winding drum. This was the last steam-powered lift. The tanks were immersed in the canal at the bottom but at the top they were pushed against hydraulically sealed doors by hydraulic rams. The levelling of the incline for the top tank was to counteract the reduction in load as the bottom tank became buoyant.

The plane saved 230m³ of water per passage, 90 per cent of what was normally used, and reduced passage time considerably; from 70 minutes to 12 minutes. However, it was uneconomic because there was not enough traffic and it closed in 1910 after just a decade of use. The plane's boiler room now contains the Foxton Canal Museum, including a working model of the plane. The locks, lock keeper's cottage and junction distance post are all Grade II listed. At the bottom of the flight is Foxton Junction with

Continues on page 92.

a horse-drawn trip boat using horse and crew in traditional dress. The Foxton Locks Inn can be very busy with visitors. It was the first to be taken into ownership by a British Waterways partnership but is now back in private hands.

The Market Harborough Branch was originally the main line, with traffic including coal and animal bones for a glue factory. It passes Gartree prison and Foxton church, which is built on a pre-Conquest site and has a Norman font and Saxon remains.

One change on the older part of the canal is that bridges have names as well as numbers, useful in this rural area. One of the first carries the Leicestershire Round footpath up towards Gumley, where King Æthelbald of Mercia held council in 749. Gumley Hall has an Italianate tower. The canal has been designated a Site of Special Scientific Interest all the way from Debdale Wharf to Kilby Bridge.

Saddington Tunnel is the shortest of the canal's three at 800m, short enough for the tunnel profile to be visible all the way through even without lights. It was built crooked, although not enough to affect the line of sight. There were slips in the approach cuttings during construction, mounds of soil from these being left near the canal to contrast with the ridge and furrow fields. The horse path over the tunnel is the most direct and obvious of them all. The tunnel acts as a bat roost and has a ghost, Anna, a headless woman.

Once the canal comes out of the cutting, there are extensive views back past Fleckney. Powerlines cross over, converging on a substation on the edge of Kibworth Beauchamp. At the time of the Domesday survey, Chiburde was the farmstead of Ciber or Cybba

and de Beauchamp was the name of the family who were to act as the king's chief panteler for centuries, a combination of dresser, butler and banner carrier.

From Kibworth Bridge, the canal drops down through Kibworth Bottom Lock and Crane's Lock to cross the River Sence on a brick aqueduct, a river which the canal is to follow to

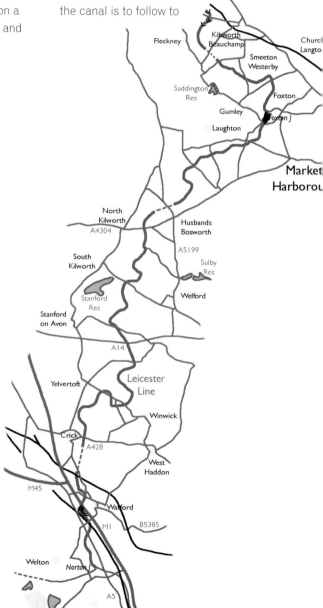

Watford Flight tranquillity beside the M1.

the River Soar. As it does so, the St Pancras to Sheffield railway line arrives alongside and is to follow closely to Wigston Harcourt. Both keep to the right side of the valley, giving extensive views over the sheep pastures below. On the far side of the valley is Wistow with its Jacobean hall rebuilt in the 19th century, 18th century church with Norman work, model railway, model village and garden centre. Wistan was the Saxon heir to the throne of Mercia, who opposed the incestuous marriage of Britfardus, his godfather, to his mother, Elfleda. Britfardus spilt Wistan's skull in order to improve his own chances of getting to the throne. A shaft of light appeared above Wistan's grave and human hair is reported to grow from the ground for an hour on 1 June where he was killed. Britfardus went insane and Wistan became a saint.

A brick church of 1834 with a 13th century stone tower by the canal gives notice of arrival at Newton Harcourt, with its three locks, Wain Bridge, ridge and furrow system and displays of dog roses. This quiet spot was the scene of rioting led by militant navvies during the canal's construction in 1795, resulting in the cavalry being called in to restore order.

The descent continues past Kilby Bridge, Double Rail and Irving's Locks to South Wigston,

The site of the inclined plane.

where the character of the canal changes. After a mostly rural run all the way from the main line near Daventry, the canal now comes within reach of increasingly urban areas.

A Rugby to Leicester railway used to cross here and head towards the 14th century church. A totem pole stands in a picnic area before Bush Lock and some of the gardens are well decorated, notably one with wrought-iron railings

Hardly a building can be seen from rural Tythorn Hill.

and holders with church-sized candles beside the canal. Frogs are oblivious.

The B582 crosses Little Glen Bridge and then both road and canal are crossed by the Leicester to Nuneaton railway. Near Little Glen Lock, 6th century grave ornaments were found in a Saxon cemetery. Across the valley, Blaby's church is merely 14th century with a fine 18th century gallery. Blaby Bridge carries the A426 over before Whetstone Lane Lock and the line of another former Rugby to Leicester railway.

Beyond Gees and Blue Bank Locks an aerial to the left marks Leicestershire Police headquarters before reaching the A563, Leicester's outer ring road. King's Lock is followed by the canal joining the river with its much more winding course. Just after the confluence the river is crossed by the former railway line, which has now become the Great Central Way cycle route where Aylestone Bridge also takes a minor road over, and travels past a church with a 1930 stained glass window.

Leicestershire county cricket ground lies to

the east of Aylestone Mill Lock. To the east of St Mary's Mill Lock are Aylestone Gas Works, now the first gas museum, including a 1920s all-gas kitchen complete with gas radio.

Leicester – named after Llyr or Lear, king of the sea, meaning darkness and death, or after the Legra, the Celtic name for the Soar – was the Roman town of Ratae Coritanorum by 100. Raw Dykes was a section of aqueduct supplying the town. Indeed, the town began as an Iron Age settlement. Later it was to be where Simon de Montfort called the first Parliament in 1265.

At New or Freeman's Meadow Lock the river comes in and crosses the canal to reach a long weir guarded by a boom before the two watercourses combine below the lock. Overlooking the lock is Leicester City's King Power football stadium of 2002.

The river leaves to the left as the canal enters the Mile Straight, a grand water avenue through the city centre, designed in 1890 for flood defence purposes but now forming an important part of the city. Old and new bridges across include some ornately decorated Victorian structures. Surrounding buildings feature the same blend of the traditional and the modern. These include the Liberty building of 1921 with a Statue of Liberty on one corner, and the De Montfort University and the New Walk Museum & Art Gallery, which has a Victorian street scene, period costumes, clocks and clockmaker's workshop, and a history of the hosiery, costume and lace industries. Leicester Castle ruins stand on a motte near St Mary de Castro church, founded in 1107. St Martin's cathedral was a parish church until 1929, originally Norman, with a 13–14th century interior and mostly 19th century exterior, a high steeple and a notable south porch. It stands next to the 14–15th century guildhall with cells and gibbet irons, fine oak panelling and a massive carved chimney piece of 1637, the townhall until 1876. The covered market is Europe's largest. Together with the Holiday Inn

these are approached along the A47 from West Bridge, which is also on the line
of the Fosse Way, the Roman road from Exeter to Lincoln. The Jewry Wall is the
largest surviving Roman building in England. It was erected in 125–130, is 7.3m
high and is part of a large public baths, now with a museum and two in-situ
mosaics. Also of note are the Jain Centre Temple & Museum, the best Indian
traditional architecture and first Jain temple in the Western world. The covered
market is claimed to be the largest in Europe.

The initials UN on a coping stone stand for Union Navigation and mark
where the Leicestershire & Northamptonshire Union Canal joined the Leicester
Navigation. The river rejoins briefly.

Coots dabble among lilies in front of a dyers' factory and swans nest with a
controlling view of Evans' Weir, where the river loops away again. At the end

The start of the Mile Straight running through the heart of Leicester.

Northbridge Tavern. A large drain discharges through the bridge abutment, dark but large enough for a small boat to enter.

Canal Carriers Wharves follow to the A5131 St Margaret's Way, which leads to St Margaret's church, a 13–15th century building with a notable Perpendicular tower and chancel. Abbey Park, with its miniature railway and boating lake, is an important green area in the city and one of England's most beautiful, opened in 1882 on what had been a marsh. It has a statue of Cardinal Wolsey, who died in the city in 1530. Towards its eastern end is the Belgrave Gate Wharf arm above Limekiln Lock and there is another arm between the lock and the next road. Again, many coconuts may be found in the canal. The number of alphabets in use on the local college's welcome board are a reflection of Leicester's large Asian population. Local sons include Engelbert Humperdinck.

In the opposite direction the road leads to the National Space Science Centre, rising like a large silver inflatable structure rather than the streamlined tower that might be expected and including a Soyuz spacecraft. The Museum of Science & Technology in the Abbey Pumping Station features canal models, four working beam engines and a history of the development of transport and power in the hosiery industry.

Further locks come at Birstall and at Thurmaston, site of a large Anglo-Saxon pagan cemetery, before the river leaves. The canal enters Watermead Country Park, an area of digging from which gravel was extracted and carried by canal to Syston until the 1980s. In the park is a large plastic mammoth, recalling that the remains of a real one were excavated here.

The Hope & Anchor public house proves popular with everyone, before the A46. The canal joins the River Wreake, at one time the Melton Mowbray Navigation and currently being

of the weir was the terminus of the Leicester & Swannington Railway, one of the first standard-gauge lines in the Midlands. Built in 1832 by Robert Stephenson, it allowed Leicestershire coal to compete with Derbyshire and Nottinghamshire and took traffic from the canal downstream. An 1834 bridge for horse-drawn wagons to cross the river is now in Snibston Discovery Park.

The canal returns to the river to allow excess water a second chance to discharge over Hitchcock's Weir before leaving for North Lock. The A50 crosses between the lock and the

Sileby Lock with its two weirs and large weirpool.

restored, which it follows back through Junction Lock to the River Soar at Cossington Lock. This detour was required in order that the canal should not be visible from Wanlip Hall. The 16th century corn and paper mill at Cossington Lock is now a restaurant. Once again, the navigation has to cross the river above a weir to reach the lock cut.

Rothley was the site of a Roman villa and has a granite Norman church with a Saxon cross and a 13th century chapel with the figure of a Knight Templar. An Elizabethan house in the village was the home of the Babingtons from the 15th to the 19th centuries and the birthplace of historian Lord Macaulay. Today it is a hotel. Rothley was also where Helen Locke lived in Freda Warrington's *The Rainbow Gate*. Rothley Brook joins above the sewage works and the river is followed by the A6 to Loughborough. Across the fields, Cossington's church has excellent Victorian stained glass.

At Sileby, the vicar hands out oranges to local schoolchildren on Whitsunday in memory of the victory at Waterloo. A ring of ten bells serves as a reminder of the presence of the large St Mary's church the rest of the year.

At the downstream end of Mountsorrel Lock it is necessary for portable craft to relaunch on the far side of the road that crosses but the wall of the Waterside Inn and the humped bridge make

the approach of traffic blind from both directions. Canal navvies were housed in huts at the Navins.

The longest brick arch bridge in Europe, carrying a conveyor, and the Railway Inn remain as reminders that a private railway link of 1860 undermined the viability of Mountsorrel Wharf, which was built in order to load pink granite from the quarry, that used to coat the neighbourhood in dust. Mountsorrel has a covered market cross and the Stonehurst Family Farm & Motor Museum.

The canal passes back under the A6 and rejoins the river before turning sharply left to follow the St Pancras to Sheffield railway again for the rest of the way.

Approaching Barrow upon Soar, there are extensive views over the meadows to the left. This time it is the river that makes the loop while the canal takes the more usual straight line, leaving at the Navigation Inn. Superior housing surrounds lawns sloping down to the canal with the occasional dovecote or tree house.

Barrow upon Soar was the site of a Roman settlement and later limestone quarries. Beyond Barrow-on-Soar Deep Lock, the canal meets the river again below a restaurant and the Soar Bridge Inn, the latter making reference to the venerable red brick bridge across the river. Beyond the inn are more arches, this time supporting the railway as part of a retaining wall built into the side away from the river.

Pilling's Flood Lock is usually left open as the canal leaves the river at the foot of Catsick Hill. The canal crosses Loughborough Moors at a level higher than that of the surrounding land, giving extensive views south-west to Charnwood Forest and northwards over the Soar valley.

An anti-aircraft gun at the front of TS *Venomous*, devastated by a fire in 2012, gives notice of arrival in Loughborough, named after the Old English man Luhdede. Beyond it is the terminus of the Great Central Railway, Britain's only remaining mainline steam railway, running for 13km from Birstall. There is a museum and historic locomotive collection here. The line was used for filming the 1993 Richard Attenborough-directed film, *Shadowlands*. The bridge that extended the line to the current freight line to Nottingham still stands across the canal among the derelict factory buildings.

One factory still active is 3M's beauty products mill next to the A60, which crosses on the Duke of York Bridge and leads towards the John Taylor Bellfoundry Museum. This is situated in the world's largest bell foundry complex, the products of which have included the Great Paul bell for St Paul's Cathederal in London. A tower of 1922 was Britain's first grand carillon. It is 46m high and now houses an armed forces museum. All Saints church has an exceptional 15th century roof and clerestory and a magnificent and unusual Somerset tower.

The derelict factories appear to not bode well for Loughborough but the atmosphere improves and the authorities really make maximum use of the canal as an asset as it curves round the town past the Boat Inn, recent housing and a playground to enter from the north-east. A branch of the canal leads back to the south-east to serve Loughborough Basin. Just before the junction is the Chain Bridge, the predecessor of which used to have a chain across at night to prevent boats slipping through in the dark without paying tolls.

The Hope & Anchor wharf – busy on a sunny spring day.

A Charnwood Forest branch had also been planned to carry trucks on rafts. Loughborough is a hosiery town and witnessed Luddite riots. Its November fair dates from the 13th century.

After the junction, the Albion Inn faces the canal, although this was not always the case, since at the time at which it was built it was a farmhouse facing the other way.

There are two more locks, Loughborough and Bishop Meadow, before the canal rejoins the River Soar with Nottinghamshire on the far bank and a large industrial estate on the near side. From here the navigation uses the River Soar continuously to the River Trent, except for the occasional small loop.

Normanton has a chain ferry. Its buildings include a cruciform church with a 13th century steeple right next to the river, the Plough Inn, a thatched post office and summer house and a number of wooden houses that must surely be weekend haunts. It is a most interesting little village to visit.

The canal leaves under the A6006, which now has the canal on the north side and a mill stream and weir stream on the south, confusing as powered craft can use parts of various channels. The lock channel cuts a corner at Zouch, returning to the river after a lock. The road runs down the main island, serving the Rose & Crown. A large castle playhouse faces on to the canal.

Loughborough Boat Club's boathouse is passed and the river flows placidly below a steep wooded bank, unusual in this valley, towards the Devil's Elbow and the White House pub.

The large hill, which has been visible on the left for some while, carries Castle Donington and East Midlands Airport. At Kegworth the river is just 3km from the airport's touchdown runway. Kegworth includes a large house with

Railway arches support a retaining wall at Barrow-upon-Soar.

Ratcliffe Lock picked out in the grass beside its replacement.

spacious grounds, a conspicuous dovecote and a beautiful rock garden. The river and canal separate round a lock island in front of it.

Ratcliffe's church dates from the 13th century and has whitewashed walls but it is totally overwhelmed by the eight cooling towers of the 2GW Ratcliffe-on-Soar power station, plumes of water vapour climbing skywards to make it more conspicuous over a great distance, if not improving the visibility for pilots. It was targetted by environmental protesters. Warm cooling water is pumped into the river. This is the source of the various sets of powerlines that have been crossing and following the river since Barrow upon Soar.

The River Soar finally discharges into the River Trent just above one of its massive weirs. The far bank is part of a large island with two locks so craft need to travel upstream for 300m to Trentlock to go either way on the Trent or up the Erewash Canal.

13 Erewash Canal

The Erewash Canal runs southwards from Eastwood to join the River Trent at Long Eaton. Opened in 1779, it served the Nottinghamshire/Derbyshire coalfield and its prosperity resulted in the building of a number of other canals, the latter now all closed. Bought by the Grand Union in 1932, the Erewash Canal was allowed to fall into disrepair, was brought back into use during the Second World War, was again neglected and was finally restored and reopened in 1973 after a combined effort by the Erewash Canal Preservation & Development Association, Derbyshire and Nottinghamshire County Councils and British Waterways.

The canal closely follows the small River Erewash and is in turn followed by the railway as far as Long Eaton. Indeed, it was a meeting in Eastwood early in the 19th century that led to the establishment of the Midland Railway that was to result in the downfall of many of the canals.

The Great Northern Basin at Eastwood has been restored to an attractive terminal with grassed areas, seating, a swing bridge and the Great Northern public house. The large Langley Mill stands near the A608. Two other canals join the Erewash at the basin above the first of its 15 locks. The Cromford Canal has been cut in two by the collapse of a tunnel as a result of mining subsidence. Subsidence has also resulted in the closure of the Nottingham Canal, the line of which closely followed the Erewash Canal all the way to Trowell.

Recent deep pits on the left are followed by works on both sides before the canal breaks into open country. There are extensive views over the Erewash valley on the right towards Shipley, Lawrence's Shepstone, and the scenery is generally at its best as the canal winds through copses. Bridges have flat, cast-iron decks between simple abutments, built in the dark brick of the area. Occasionally, there are collections of old lorries and other debris beside the canal. Graffiti mostly relates to angling and, although the canal is quite wide, anglers rest the tips of their long rods on the opposite bank to the one on which they are sitting.

The disused Bennerley Viaduct at Awsworth.

The Nottingham and Erewash Canals leave the Great Northern Basin at Eastwood.

Locks often have bypass channels with steep drops at the entry ends. Eastwood Lock has a plaque remembering local author DH Lawrence on the right side. Lawrence was born in New Eastwood and based so many of his books on the area. One, *The Rainbow,* describes the effects of canal construction.

The river being the county boundary, the canal moves from Nottinghamshire to Derbyshire for the rest of its route.

Shipley Lock had the Shipley Boat Inn, Shipley Gate being Lawrence's Shipley Boathouse,

and a selection of buildings that were formerly a slaughterhouse and stable block, the latter with a sagging roofline. Not all buildings are so interesting, however, and, after a marshy area, there are long lines of featureless modern houses with small windows front the canal at Cotmanhay. Having said that, Cotmanhay can boast the old Bridge Inn beside an arched bridge that had to be replaced as it settled enough to give problems to some canal craft.

The 430m long lattice trestle

Distance
19km from the Cromford Canal to the River Trent
Highlights
Bennerley Viaduct
Navigation Authority
Canal & River Trust
Canal Society
Erewash Canal Preservation & Development Association
www.erewashcanalpreservationand developmentassoc.org.uk
OS 1:50,000 Sheet
129 Nottingham & Loughborough

Contemporary buildings and an interesting roofline on the former stables by the lock at Shipley.

Bennerley Viaduct used to carry twin railway tracks right across the valley to Awsworth and is now preserved as a listed structure.

Stenson's Lock is in open country but by Common Bottom Lock the canal has become hemmed in with houses and small works. An occasional line of poplar trees adds variety and there was another bridge by the lock, now burnt out. The canal forms the eastern boundary of Ilkeston, a market and textile town with the Ilkeston Charter Fair each October. St Mary's church, distinguished by its tower, dates from 1150 and has an unusual 14th century stone screen.

Cossall, called Cossethay in *The Rainbow*, has a street of terraced houses leading down to the canal. An open playing field area with a rugby pitch and a track where children might be seen driving miniature replica vintage cars follows, and nature pushes back with giant reedmace, bulrushes and perhaps the odd vole swimming across the canal.

Pasture Lock, at Stanton Gate, still lives up to its name, with horses grazing in the meadows. These stretch away to the hill that is topped by Sandiacre church with its original Norman work and 600-year-old font, surrounded by a clump of half-timbered buildings.

A substantial-looking footbridge leads away over the railway to the larger Stapleford, which keeps its distance. The pride of the canal's factories is Springfield Mills, built in 1888 for lacemaking, fronted by a prominent chimney with intricate

The end of the former Derby Canal.

Springfield Mills retain their striking chimney.

and an electricity substation. A mill chimney by the canal has no less than three tiers of aerials around it. Below the lock, the canal is flanked by the playing fields near Trent College, by a school and a fire station and by a tree-lined urban boulevard, the playing fields continuing opposite another school. Facing the canal and its swans is the Barge Inn.

The canal turns sharply under the road by the 15th Long Eaton Sea Scout & Guide Group's premises. It again bends left by a public house with a canalside garden.

brickwork at the top. It has been converted for residential use. Opposite it is the Plough. The Red Lion and White Lion follow in the centre of Sandiacre, a town that makes a feature of the canal, with lawns reaching to the water, seats and flowerbeds.

Beyond the busy A52 and tucked in next to the railway is the jewel of the canal, the lock cottage at Sandiacre Lock, base for the canal society. Complete with all its stable doors, it shows rural canal architecture at its best. Just above is the end of the Derby Canal. Abandoned in 1964, it joined the Trent & Mersey Canal at Swarkestone.

The reach leading down to Long Eaton Lock has timber and pipe depots, a petrol station

The canal company's dry dock is now part of Mills Dockyard moorings. Houseboats line the bank down to Trentlock, overlooked by the garish Steamboat Inn with its funnel and ventilators on the front. The Trent Lock pub on the other bank has a landscaped children's area and public toilets in the car park at the back.

Below the lock is a complex junction with the River Trent. A navigation cut comes off at this point and the River Soar, the northern end of the Leicester arm of the Grand Union Canal, enters opposite. Above it all, aircraft descend into East Midlands airport at Castle Donington over the cooling towers of Ratcliffe-on-Soar power station.

The canal flanks an urban boulevard in Long Eaton.

14 Grand Union Canal: Aylesbury Arm

The Aylesbury Arm of the Grand Union Canal is somewhat less significant than it was when originally planned. It was to have been the Western Junction Canal, running to Abingdon where it would have been continued by the Wilts & Berks Canal to Melksham and then the Kennet & Avon Canal to Bristol. Abingdon already had the River Thames from London and, not far north, the Oxford Canal towards Birmingham. As it happened, the Western Junction never got beyond Aylesbury and Abingdon missed out on becoming a major transport interchange, as it was to do again later when it rejected the Great Western Railway.

A contributory problem was the starting point at Marsworth, just seven locks down from the summit level of the Grand Union Canal. Because of the inevitable water supply problem and the fear that it would drain the main line, all locks on

On the Marsworth Flight near Wilstone.

the Aylesbury Arm were built as narrow beam and this would have meant that only narrow beam craft would have been able to work through to Bristol. As it was, Wilstone Reservoir, which supplies the Aylesbury Arm as well as being pumped to the main line summit level by pumps at Tringford, had to be enlarged twice. Like the Tring Reservoirs, it sits between the Aylesbury and Wendover Arms, all doubling as nature reserves with breeding and migratory birds and a small heronry.

Nevertheless, the idea had popular support from the people of Aylesbury and the Marquis of Buckingham, and the Aylesbury Arm was opened in 1815, remaining commercial until the 1950s, after which it became semi-derelict. In its heyday it took agricultural produce and livestock out of Aylesbury and delivered building materials, timber and coal in return.

Now fully restored, it poses a dilemma for the narrowboat user. The Aylesbury Arm is one of the quietest and most rural canals in the country, a charming find within easy reach of London, yet it has 16 locks on the descent to Aylesbury and 32 locks for the 20km return trip on this dead-end canal.

The Aylesbury Arm leaves the main line at Marsworth Junction, 300m north-west of the Lower Icknield Way.

The first eight locks form the Marsworth Flight. They average 200m apart. The first pair come as a staircase. Many of the locks have houses alongside; not with traditional lock keeper's cottage styling but their locations cannot be accidental in this rural landscape.

The line is almost due west across Buckinghamshire although sections of Hertfordshire are intersected in the early part of the route.

The fourth lock, which has a wall of periwinkle at its lower end, swings the canal on to the straight line it is to follow for the next 3km. Looking back,

Distance
10km from Marsworth Junction to Aylesbury
Highlights
The highest point in the Chilterns
Watching your own arrival
Navigation Authority
Canal & River Trust
Canal Society
Aylesbury Canal Society
www.aylesburycanal.org.uk
OS 1:50,000 Sheet
165 Aylesbury & Leighton Buzzard

the line points to the tower of Marsworth church. Several of the fields thus far have held horses but the land is now predominantly laid down to crops.

As the last of the eight locks is approached at Wilstone, the only canalside community, the pristine fields of bright-yellow oilseed rape give way suddenly to an apparently ploughed field. From beyond the hawthorn hedge come the satisfied grunts and squeals of a herd of pigs wallowing in a pool just out of sight but not out of range of sound or smell.

From here the gradient eases as the canal distances itself from the foot of the Chilterns. This Vale of Aylesbury countryside is flat and unspoilt, with the traditional English open-farming character. To the south the scarp slope of the Chilterns rises steeply to the beechwood-clad Coombe Hill, the highest point in the Chilterns standing at 260m, 170m above the vale

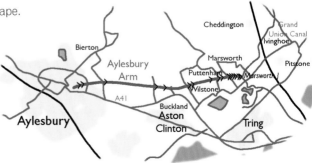

Happiness is pig-shaped. The lock's neighbours at Wilstone.

here. Interestingly, 5km to the south-west is another Coombe Hill that exactly matches the 260m height of this one, the latter being topped by a monument and hiding Chequers, the country residence of the Prime Minister.

Two Puttenham locks draw attention to the village of Puttenham to the north. One of the church's vicars was Christopher Urswick. As Richard III's Recorder of London, he supplied information to the exiled Henry Tudor, later marrying him to Elizabeth, the daughter of Edward IV.

The village of Aston Clinton was developed largely by the Rothschilds who settled here in 1851 and proceeded to spend some of their great wealth in the area.

Lines of communication converge with the canal as it approaches Aylesbury. To the left lies the A41, the current form of the Roman Akeman Street, while to the right is the course of a former railway that ran from Aylesbury to Cheddington.

The land becomes very low-lying around Broughton Lock, giving extensive views, and there has been bank restoration work, sheetpiling and fabric membrane laid on top of the embankment to give stability.

The canal probably follows a former line of the Bear Brook that now runs to the south, possibly diverted to serve a mill. The brook's present course is on the side of a slope and in times of flood it overflows down towards the canal. A flood relief scheme for Aylesbury makes use of two areas of low ground as flood storage areas. The Stocklake Brook area lies further north but the Bear

Brook flood storage area is located between Broughton Manor, the housing of Aylesbury and the canal.

There are older factories around Aylesbury Lock and Aylesbury Bottom Lock. One business worthy of note was Hazell Watson & Viney, for many years involved with the printing of Penguin Books until the company was bought by Robert Maxwell, at which point Penguin withdrew all their custom. Another was the Aylesbury Condensed Milk Company, attracted in 1870 by the canal and later owned by Nestlé.

The appearance of the town was substantially changed in the 1960s by redevelopment. One recent construction, which does deserve praise for being within the spirit of canal architecture while embodying original thinking, is the well-proportioned footbridge across the entrance to the terminal basin.

Being the bottom end of the canal, a weir is needed and takes the form of a neat semi-circular brick overflow that feeds the head of the California Brook on the south-east side, joining the Bear Brook to the west of Aylesbury. Behind the moored boats opposite is an aerial reaching out of some industrial premises. Beyond is the town centre, gathered around the chunky concrete county hall tower block.

The canal basin terminates at an office block that obscures a 13th century church but has a mirror window that reaches almost to water level so the boater can have the unusual experience of watching his own arrival. On the other hand, he can see all kinds of reflections in the angular glass edifice beyond it – not a swimming pool as might be thought but the offices of the Equitable Life Assurance Society.

Aylesbury Basin and its incongruous neighbours lie beyond the recent footbridge.

Grand Union Canal: Paddington Arm and Regent's Canal

The Grand Union Canal links Birmingham with London. At the Birmingham end it is well connected but in London the through route leads to the River Thames, convenient for the docks but too low for the industries on the higher river terraces. The Paddington Arm was built as a lock-free pound to link the Grand Union Canal with the north-west of London, terminating at Paddington but subsequently being extended via the Regent's Canal to the docks.

In the days before railways, canals were an efficient form of transport and there is a record of troops taking seven days to travel from Paddington to Liverpool by boat in 1806, much quicker than marching and arrived the troops much fresher.

Bull's Bridge Junction is on the main line of the Grand Union Canal, crossed by a Grade II arched brick bridge dating from 1801, with stop gates underneath. On the left is a 19th century toll house, the offices of the Colne Valley Passenger Boat Service.

Victorian terraced houses back directly on to the canal at West Kilburn. Half Penny Bridge crosses beyond.

The former Kensal Green Gasworks.

Distance
22km from Bull's Bridge to Paddington Basin
14km from the Paddington Arm to the River Thames
Highlights
Little Venice
Paddington Basin
Regent's Park and London Zoo
Camden Lock Centre
London Canal Museum
Limehouse Basin
Navigation Authority
Canal & River Trust
OS 1:50,000 Sheet
176 West London
177 East London

The Penzance to Paddington railway crosses and is never too far from the canal for the rest of its journey, Brunel having chosen to locate the terminus of his Great Western Railway next to the canal basin at Paddington.

Each side of the railway are car depots, the one on the offside being surrounded by masses of barbed wire and causing an enormous traffic jam on the site of the Southall Gasworks, now represented only by a massive gas holder. It has been used in filming *Doctor Who* and *The Sweeney*. The Mitre Dock and Ticklers Dock belonged to Ticklers, who made jam and supplied the army during the First World War. The last commercial users of the canal were biscuit-makers Kearly & Tonge, who received coal deliveries here.

Among the dumped debris, which the blackberry bushes and other vegetation attempt to disguise, is an overflow weir feeding Yeading Brook. By the A4020 is the Hambrough Tavern, which was burnt down in 1981 race riots between Asian youths and the police.

The extensive 19th century Hayes brickfields have been landscaped and the canal arms filled in. The arched Spikes footbridge crosses, moorhens nest and herons fly over. A pocket of tranquillity has returned. A new development is located around Engineer's Wharf Moorings. During the Second World War the canal was used for testing lifeboats at Greenford.

The 14th century St Mary's church was home to Nicholas Ridley, a vicar who was burnt at the stake in Oxford in 1555 after supporting Lady Jane Grey. One of the sharpest bends on the canal is by the Black Horse public house, an area of interesting aromas; fresh coffee wafts from the Lyons Tetley works, to be replaced by fresh bread from the Mother's Pride bakeries and, no doubt, the Guinness plant could also add to the aromas.

After this, the canal breaks into its most rural section. A laminated wooden arch bridge leads towards Horsenden Hill, which rises to 84m on the left bank with rugby pitches. After a large wooden owl, Sudbury Golf Club is located around its lower parts. Opposite is Perivale Wood, established as a nature reserve in 1904, a remainder of the ancient Middlesex Forest.

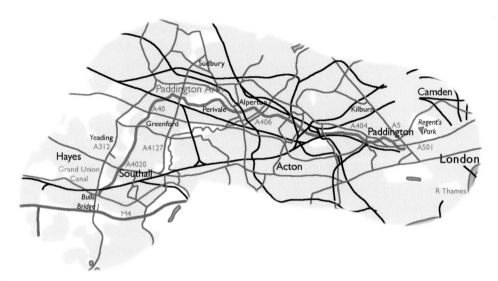

Houseboats are moored alongside the canal as it approaches the Paddington Stop.

Acton Lane Bridge is colourful. The Grand Junction Arms next to it includes a three-dimensional mock-up of a working canal boat and a decor of roses, castles and lace-edged plates.

The bridges are completed with a complex of railway arches, more following at Old Oak Common. Here, little is seen except a tall aerial and some old factories on the left bank and open space behind a long low wall on the right. Beyond the wall and rather lower than the canal are the massive railway yards that act as a holding area for engines and coaches within easy reach of

Paddington and will service Crossrail trains. Across the railway the North Pole depot has moved from servicing Eurostar to Western electric express trains. Soon the Western main line comes right alongside, as is shown by two spans of bridge, the smaller one crossing the canal while the larger one on the right is over the railway. This is where Brunel tried out his atmospheric railway system in 1840 before using it unsuccessfully on the London & Croydon Railway and in Devon.

Kensal Green cemetery on the left bank was opened in 1832. Its 23ha include three royal mausolea. Isambard Kingdom Brunel, William Thackeray, Anthony Trollope, Wilkie Collins, Decimus Burton and Emile Blondin are among those interred within its walls. Water gates opening on to the towpath allowed hearse barges

Paddington Basin is surrounded by high-rise buildings.

to be used, a most dignified way to make a departure, in all senses of the word.

Port-a-Bella Dock was built in 1894 for rubbish barges. These days it has a Sunday market, craft and canalside centre, offers evening canal cruises and has been used by the BBC for filming *The Apprentice*.

Carlton Bridge, with its cast-iron balustrades, was built in 1870. Exactly a century later came the elevated A40 Westway, sweeping out above the canal but not crossing at this point. The western portal of the Crossrail tunnel is adjacent.

The canal warden's hut is incongruous in front of the Warwick Estate tower blocks. The church of St Mary Magdalene is just incongruous. Built between 1868 and 1878, it is in Victorian Gothic style and features a prominent red-and-white-striped spire. A wall is brightened up by a colourful mural made from debris collected by Stowe Youth Club with artist Kevin Herlihy.

The Paddington Stop public house is on the straight that leads to the former Paddington Stop toll house, the bridge built at the turn of the last century, and some public toilets. Beyond the bridge is Little Venice or Browning's Pool, one of the best known canal sights in London. Opened in 1801, the basin includes Rat Island or Browning's Island, an area of weeping willow trees, popular with mallards, tufted ducks, Canada geese, moorhens, coots and swans and named after Robert Browning, who lived from 1862 to 1887 in Warwick Crescent in a house that has since been demolished.

Little Venice is one of the most colourful canal scenes in London, an urban waterway at its most classy. Jason's Trip operates from here, inaugurated in 1951 for the Festival of Britain, and Jenny Wren Canal Cruises operate to Camden along the Regent's Canal, which leaves the northern end of the basin. Waterbus services began from here in 1959.

The canal continues through a park and beneath the Westway, the A40 on the top deck and the A404 Harrow Road underneath. The A4206 Bishop's Road Bridge upgrade involved jacking the old bridge up 10m while its replacement was built underneath, in the process removing an early Brunel bridge for relocation over the canal. Craft are moored in front of Brunel's railway terminus for south-west England and Wales. Agatha Christie and Arthur Conan Doyle were among its repeated users in literature. Indeed, Dr Watson's surgery was near the station in the Sherlock Holmes stories.

One old building has crane brackets and other ironwork bolted to it to help unload barges, while other buildings are recent. A modern cable-stayed footbridge crosses Paddington Waterside near St Mary's Hospital where Alexander Fleming discovered penicillin in 1828 – his laboratory now a museum – and there is a stained glass window dedicated to him in St James' church.

The Regent's Canal, named after the future George IV, was completed in 1820, running through open country to the north of London to link the Paddington Arm of the Grand Union Canal with the docks. It was part of John Nash's metropolitan improvements of the Regency, the only coordinated building plan in the capital's history, a scheme that included Regent's Park, St James' Park, Regent Street, Trafalgar Square, the Strand and the Suffolk Street areas. The canal carried much coal, timber, building materials and food produce, serving many factories along its length, but never made the anticipated profit because of competition from the railways. Indeed, the canal's owners attempted

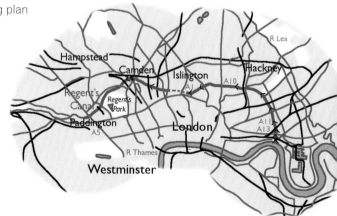

The Regent's Canal from Little Venice with Browning's
Island on the right.

to turn it into a railway in 1845. It was the ceasing
of munitions traffic after 1945 that really sent the
canal into decline although working narrowboats
and horse- and tractor-drawn dumb lighters
continued until the early 1960s and there is still
some trade, mostly in timber.

Now, however, the canal's primary function is
as one of the most valuable amenity waterways
in Britain as it provides a green ribbon around the
heart of this built-up area of London.

The Regent's Canal leaves the Paddington
Arm at Little Venice in Paddington, base for
the May Canalway Cavalcade, founded in 1983.
Leading east from the basin on each side of
the canal are plane tree-lined avenues of fine
Regency houses, as well as Junction Cottage, a
canalside house.

This reach continues to Maida Hill Tunnel,
which is only 249m long but has no towpath.
It is difficult to leave the canal because of the
continuous spiked railings.

Passing over the straight tunnel is the equally
straight A5, originally built as the Roman Watling

Street from London to St Albans. Spoil from the
tunnel was spread on the field of local landowner
William Lord. This field, opened in 1809, now
forms the home of Marylebone Cricket Club,
which represents England internationally; Lord's
is one of the world's best-known cricket grounds,
and features a museum containing the Ashes of
a burnt bail.

The canal skirts the northern edge of Regent's
Park. Formerly Marylebone Park, which was
used as a royal hunting ground by Henry VIII,
who seized it from the Abbess of Barking in
1538, until the time of Cromwell, who cleared
the trees for farming. The 1.7km² park was laid
out in 1812–1827 by John Nash and opened in
1838, being named after the Prince Regent. The
lake was used in the filming of *Brief Encounter*.
Overlooking the park is James Bond's eight-
storey Secret Service office block.

There are several interesting bridges along
this reach. One carries the Tyburn in conduit.
Another is Macclesfield Bridge, named after
the Earl of Macclesfield, the Regent's Canal
Company's first chairman. With three segmental
brick arches and circular spandrel holes, it is

Macclesfield Bridge, also known as Blow Up Bridge.

also known as Blow Up Bridge, after an explosion that took place in 1874. The narrowboat *Tilbury*, loaded with gunpowder and benzol, was one of three being towed by a steam tug when it exploded as it passed under the bridge, reducing it to rubble, damaging nearby houses and killing three crew members, whoes bodies were never recovered. The bridge was subsequently rebuilt, its cast-iron Doric columns being rotated to put the rope-incised sides away from the water. The bridge stands today, its cast-iron capstones proclaiming that they were cast at Coalbrookdale.

The park is best known as the home of London Zoo, one of the world's largest at 15ha and also one of the oldest. The zoo was laid out in 1827 by Decimus Burton and carries out much important research; the public were not admitted until 1847. It houses 8,000 species of animal, bird, reptile and fish, together with a bar, restaurants and cafés. A waterbus landing stage serves the zoo, the canal passing by the antelope terraces from where Arabian oryx gaze at the boater and across at Lord Snowdon's dramatic aluminium Northern Aviary of 1965.

The canal turns sharply left towards Primrose Hill, with its elegant Victorian houses and canalside gardens. Before these is St Mark's church, built beside the canal, a squat stone edifice with a slender spire. Bridges were designed by Robert Stephenson. One carries the West Coast Main Line. When it was first opened, trains were hauled by cable from Euston up the steep slope over the canal.

The Pirate Club was formed in 1966 to promote watersports. The building is designed as a pirate castle complete with drawbridge and battlements, club room, canteen, mini gym and workshops; it was opened in 1977. It also has a narrowboat, named the *Pirate Princess* in 1982 by Prince Charles, who demonstrated that narrowboat handling was not one of his finer attributes, the tiller being seized from the future monarch's grasp at one point by the centre warden. The club membership peaked at 1,200 and fundraising in the early days included pirate raids on commercial craft such as the *Jenny Wren*, staff trying to get their hands on the booty needed for development of the centre before it found its way into young pirate pockets.

The former Dingwalls Timber Wharf & Dock has become the Camden Lock Centre, opened in 1973. The original stables, with hay lofts above, have become craft workshops, restaurants, a weekend market, night club and picture and antique shops. There is a cruising restaurant boat and there are open-air jazz concerts, ox roasts, Caribbean parties and exhibitions – a thriving hive of activity.

Hampstead Road Lock has a two-storey castellated lock cottage that was the first of the buildings in the area to have battlements and is now the canal information centre, reached by a cast-iron bridge over the canal. The lock marks the end of the 43km Long Level pound from the Grand Union Canal main line at Cowley and Norwood Top Lock. From here it is a 26m descent to the Thames.

The Constitution pub begins a straight reach that is bordered by a large recent housing complex and then the Jubilee Waterside Centre.

The skyline ranges from the British Telecom Tower to the Tower 42, formerly the NatWest Tower, but only one of the five St Pancras gas holders remains with its delicate 1860 tracery, a listed structure originally belonging to the Imperial Gas Light & Coke Company.

The extensive King's Cross redevelopment scheme has been taking place on both sides of the canal while the canopy of King's Cross station may be seen on the right, from where the East Coast Main Line passes under the canal and from where the Hogwarts Express used to leave from platform 9 $^3/_4$. Disused stop gates under a bridge were positioned to protect the railway from flooding in case of a well-aimed wartime bomb. The other set of stop gates are under Maiden Lane Bridge, which has fine cast-iron work dating from 1850.

Battlebridge, named after Boudica's fight with the Romans, became King's Cross. The name is retained in the Battlebridge Basin, also known as the Horsfall Basin from the name of

Lord Snowdon's Northern Aviary at London Zoo.

the local landowner at the time of the canal's construction. The 6,000m^2 basin has the London Canal Museum in a former ice-cream company ice warehouse and King's Place, London's newest concert venue.

The 878m-long Islington Tunnel with no towpath is straight with daylight at the far end. It was built using a horse-drawn railway along the invert for removing the spoil and was opened in 1816. From 1826 to the dawn of the 20th century it had a steam tug on a platform like a railway engine pulling along a chain laid along the bottom of the canal and passing round two huge drums set 2.4m apart. Its successor was said to have been the ugliest boat ever seen on British canals, a coke-fired engine that resembled a corrugated shed enveloped in steam, taking nine loaded barges through in 35 minutes. Diesel

Camden Lock, probably the busiest place on the British canal network.

tugs took over in 1930. An alternative use for the tunnel came in the cold winter of 1855 when skaters were able to use it.

Unpowered craft are not permitted to go through. The walking route is long, follows a network of streets, passes shops, most days takes in a busy market street with a large winged sculpture of tubes, and crosses the dual carriageway A1 with central-reservation fencing – not a trip for the sensitive.

The canal leaves the tunnel mouth between trees and Georgian terraces painted by Walter Sickert as the *Hanging Gardens of Islington*. Islington is from the Old English *Gisla's downland*, although nothing could be much less like downland these days.

Hackney's Victoria Park was laid out in 1842–1845 by James Pennethorne, a protégé of John Nash. It is London's oldest municipal park, the largest in the East End at 88ha, and includes a picnic area. Bonner Hall Bridge, with its fine ironwork railings, was part of the original formal entrance to the park.

Johnson's Lock is left past the Ragged Schools Museum with a Victorian schoolroom and local domestic kitchen. Dr Barnardo used it to house London's biggest ragged school, a charity that benefited from the takings earned from the lectures John MacGregor gave about his groundbreaking 19th century travels with a canoe, giving free education to the poor.

Below the basin is the Limehouse Link. At the time of building it was the most complicated and expensive British road ever; £220 million

Waterpoint, the water tower at St Pancras Lock, is a new arrival, acting as a viewpoint and art gallery.

for 1.8km. The construction period left a new marina with a link road between the City of London and Canary Wharf, the Isle of Dogs and the Royal Docks passing underneath.

The Limehouse Cut connection was opened in 1968, replacing a direct link into the Thames for that canal. Near it is the octagonal Limehouse Accumulator Tower of 1869, which stored water to power hydraulic cranes and locks. St Anne's church is a notable design by Hawkesmoor and is entitled to fly the White Ensign. It had the first illuminated clock in the country, the highest church clock in London.

Limehouse Lock was the largest canal lock in London at 100m x 18m, allowing 3,000t ocean-going ships through to unload directly into narrowboats and lighters. However, it was only operated for three hours before high tide between 6am and 10pm to avoid draining the basin. The traffic moved downstream in the 1960s. This lock has been replaced by a smaller lock, just 30m x 8m, needing only a tenth of the water of its predecessor. It has computer-controlled sector gates. Landing is difficult, as is launching on to the River Thames on the far side of the lock for those not locking through. A path leads up the side of the swing bridge, opposite which is the Gordon Ramsay Narrow gastropub in the former dockmaster's house, from where riverbuses can be seen racing along the river between the City and the Docklands developments and helicopters fly over to the City airport.

Limehouse Lock leads out on to the tidal Thames.

Basingstoke Canal

The canal was completed in 1794 to link the market town of Basingstoke with the capital via the Wey and Thames. With the coming of the railways the canal traffic declined and the canal company went into liquidation in 1866, although commercial traffic used it until 1949.

In 1966 the then Surrey & Hampshire Canal Society was formed and reopened the canal in 1991, run by the Basingstoke Canal Authority for the county councils. Surrey County Council supported the scheme with enthusiasm and even Hampshire County Council maintain the towpaths well and have a depot and several maintenance craft on the canal. There are claims that it is the most beautiful canal in the country. Certainly it is one of the most well-wooded.

The 1.1km Greywell Tunnel was passable into the 1950s but part of it has collapsed and the western end has been fenced off. It contains an estimated 12,500 bats and is the largest bat roost in Britain, including the world's second-

King John's Odiham Castle.

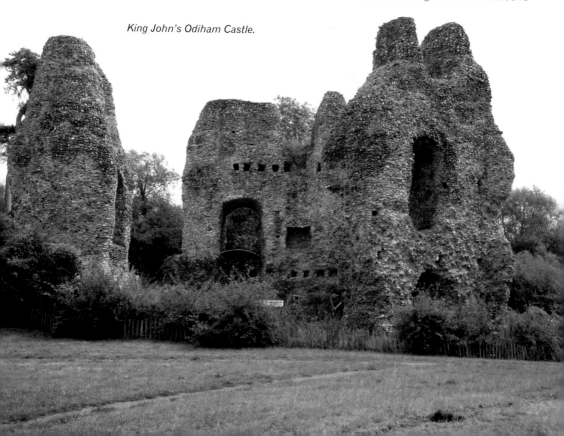

largest population of Natterer's bats. For this reason permission has not been granted for the tunnel to be restored for use by boats. There has been talk of building a second tunnel but there is no guarantee that bats would not move into that, too. The walking route is poorly marked.

The other end of the tunnel is amazingly different. A footpath leads to the tunnelmouth, the portal of which was rebuilt in 1975. As with all the bridges on the canal in Hampshire, a location plaque is fixed to the downstream face. Floating weed covers the canal surface but the water is absolutely clear and the canal is wide.

The setting of the first part of the open canal is a rural one, alternating between open meadows and woodland.

Just after the River Whitewater, the remains of the three-storey octagonal flint keep of Odiham Castle, built midway between Windsor and Winchester from 1207 to 1210 for King John, stand among the trees on the left. The castle once had two rectangular baileys, earthworks and a moat with a small outer enclosure, also moated. It was from here that King John set out to sign the Magna Carta in 1216. The French took it in 1216 during the first Barons' War. David II was imprisoned from 1346 to 1357. It became a 15th century hunting lodge and was ruined by the 17th century but partially restored in 2008.

A boom then reaches most of the way across the canal upstream of the first winding hole. From here the water becomes murkier as the weeds thin, both results of usage by boats hired out from a centre at Colt Hill. After North Warnborough the canal is followed by the Odiham bypass, which can be heard but not seen on the left.

The Barley Mow at Winchfield is the next port of call for the thirsty brigade. On a summer evening the click of leather on willow and sporadic applause break the silence from somewhere beyond the towpath hedgerow. A small swing bridge crosses.

Attractive riparian properties are met.

Between Crookham Village and Fleet is the Fox & Hounds. Houses back onto the canal at Fleet and shops lie to the north of Reading Road Bridge. In a Halifax survey, the Hart area, centred here, was the best place in the country to live.

From Fleet, the canal cuts out in to wooded heathland. Norris Bridge is a high concrete structure and, on the left, leads up to the National Gas Turbine Establishment but the noise of jets can be deceptive. After another kilometre, on the left bank where the trees clear,

Distance
56km from Old Basing to Woodham Junction
Highlights
Odiham Castle
Navigation Authority
Basingstoke Canal Authority
Canal Society
Basingstoke Canal Society
www.basingstoke-canal.org.uk
OS 1:50,000 Sheets
(175 Reading & Windsor)
176 West London
(185 Winchester & Basingstoke)
186 Aldershot & Guildford
(187 Dorking & Reigate)

fencing blocks a view straight down the main runway at Farnborough airfield, the venue for the Farnborough Air Show. A variety of aircraft from the Royal Aircraft Establishment come and go.

All is not noisy, however. From here to Wharf Bridge, several water areas on the south side of the canal are nature reserves and boaters are requested to keep out. Beyond Wharf Bridge landing is not allowed on the south bank, either, but here it is because the land belongs to the army. Ranges and barracks front the canal. Red flags fly on the other side of the towpath when ranges are in use. This is Aldershot, Hardy's Quartershot, traditionally home of the army, although their presence is diminishing.

Kiln Bridge and the Tuu restaurant at St Johns.

The Frimley Lodge Miniature Railway, on the left of the canal, has extensive narrow-gauge trackwork and can raise a number of passenger-carrying trains pulled by engines from diesels to narrow-gauge steam engines.

As the canal turns east again, there is a boarded-off bridge on the left, which should give access to a large lake. A boathouse is on the left bank at Deepcut. A wall was built by the railway company near Curzon Bridge to stop locomotives from frightening working barge tow horses.

The locks of the Brookwood Flight follow. Just the other side of the railway from Brookwood Bridge lies Brookwood cemetery and its war graves, the largest cemetery in the UK, covering 2km². Its station received funeral trains.

The Tuu Restaurant Woking next to Kiln Bridge at St Johns offers Thai, Beijing and Vietnamese cuisine. The Bridge Barn has plenty of outside seating in the summer. The tower blocks of Woking are approached and new buildings beside the canal include a large Debenham's and the New Victoria Theatre.

On the right, just beyond the A320, is a children's adventure playground with a selection of skateboarding half pipes. The trees on the other bank hide a different kind of airfield from the one at Farnborough.

Wooded heathland surrounds the Deepcut Flight.

It was here on Horsell Common that HG Wells would have us believe, in *The War of the Worlds*, that the Martians landed in their invasion of Earth, watched from the protaganist's home on Maybury Hill. These days planes flying into Gatwick are more likely to be seen crossing the sky.

The canal is now firmly in the stockbroker belt. Houses on the canal banks are well hidden by greenery, however, and do not intrude. There are small boats in smart gardens that back on to the canal.

The canal's lower locks may only be used during the middle six hours of the day as a water conservation measure. From the six locks of the Woodham Flight is an almost continuous line of houseboats, mainly converted narrowboats but some obviously floating houses. After a selection of more recent bridges, the last bridge on the canal is a rounded brick arch, typical of the ones higher up. From the last lock there is a final straight to the River Wey Navigation at Woodham Junction, backed by the M25 viaduct. Beyond that are the remains of the Brooklands race track with its banked bends still intact in places. It was 4.5km long, the fastest circuit in Britain from 1907 to 1939 with a record average lap speed of 200km/h, hosted the British Grand Prix and was used for setting the last track world speed record – Guinness's 215km/h in a Sunbeam in 1922. It also had an airfield, aircraft production displacing motor racing in 1939. Today some of the classics machines of car racing and flying are on display.

17 Kennet & Avon Canal

Proposals for what turned out to be the only east–west canal across southern England began in 1660 as a broad canal, the Western Canal. This turned out to be a successful canal, no thanks to the residents of Reading, where the mob, led by the mayor, attacked construction work, which they feared would cost them their road business. John Rennie surveyed the canal route at the age of 29 and it was completed to his design in 1810, after 16 years of construction. Highly profitable for the first 30 years, it carried coal, iron, stone, slate, timber and agricultural products but went into decline in 1841 after the arrival of the Great Western Railway (GWR), which follows the line very closely to Pewsey. The GWR bought the canal in 1852 and allowed it to deteriorate. During the First World War it was used for troop training. The closure Act of 1952 was strongly resisted, the annual Devizes to Westminster Race helping to draw attention to the canal and ultimately being a significant factor in its restoration – canoeing's major contribution to our canal network. In 1956 Commander Wray Bliss took a 20,000-signature petition for restoration by canoe and cruiser to Westminster. The Kennet & Avon Canal Trust was formed in 1962 and tackled the country's biggest restoration project successfully, the canal being reopened by the Queen in 1990. Much of the towpath is used as a Sustrans cycle route.

Sheffield Lock, a scallop-sided lock, formerly turf-sided.

Reading was successively an Iron Age then a Roman then a Saxon settlement situated on the banks of the River Kennet, rather than on the River Thames, which it joins just downstream. Indeed, the Environment Agency are the navigation authority for Blake's Lock, the bottom lock on the river.

The Abbey or Forbury Loop takes in several important buildings. Prudential's offices are on the site used by Huntley & Palmer from 1822 to 1970, the world's biggest biscuit company, breaking fewer biscuits by using water transport. Only one of the original factory buildings remains at the east end.

The Scottish baronial-style prison is where Oscar Wilde was imprisoned in 1896/7 for gross indecency, writing *De Profundis* here and *The Ballad of Reading Gaol* after his release. Toad of Toad Hall also served time here.

Next to it are the remains of the Norman Reading Abbey, founded in 1121 by Henry I, who presented the abbey with the hand of St James that attracted pilgrims, and who was buried here in 1135, as was the Empress Maud, his daughter Matilda. The abbey was built with stone from Caen and from the Roman site at Silchester. King Stephen built a castle inside the grounds in 1150 but that was soon destroyed. One of the oldest known songs, John of Fornsete's *Sumer Is Icumen In*, was written down here in 1240. John of Gaunt was married here in 1359 and Parliament met here in 1453 and 1464. Only tall sections of wall remained after the Dissolution.

The first section of the navigation, the Brewery Gut, is the most difficult of all, the flow being fast with blind bends. Use is controlled by boaters who push traffic light buttons, but the buttons are too high to be reached from small craft and are recessed so they cannot be pushed with an object such as a paddle.

Vehicles are strongly discouraged, by humps and a very low speed limit, from using the A4, which crosses next. Brunel also wished to bring

Distance
141km from the River Thames to Hanham
Highlights
The Blue Line, comprising over 120 pillboxes and the best preserved defence line from 1940
Garston Lock, the only surviving turf-sided lock
Crofton Pumping Station
Caen Hill Flight, Britain's longest continuous lock flight – 29 locks drop the canal 72m over 3.6km
Avoncliff and Dundas Aqueducts
Bath World Heritage City
Navigation Authority
Canal & River Trust
Canal Society
Kennet & Avon Canal Trust
www.katrust.org
OS 1:50,000 Sheets
172 Bristol & Bath
173 Swindon & Devizes
174 Newbury & Wantage
175 Reading & Windsor

his GWR across here and up the Pang valley to Pangbourne but his clients insisted that he take it to the north of the town.

At Burghfield the canal picks up the Blue Line that joined from the south and was to follow to Bradford-on-Avon, a 1940 defence line in case of invasion of the south of England. There were 90 infantry and 40 larger pillboxes forming one of the best preserved defence lines, a conspicuous feature of this canal.

The Cunning Man (or fortune teller) is a popular canalside public house with a children's play area. All custom disappeared overnight in the 1980s, however, when the landlord shot an intruding stray dog with a crossbow, resulting in the brewery needing to find a hasty replacement for him.

Further acts of violence occurred at a nearby roving bridge, which was the site of riots in 1720

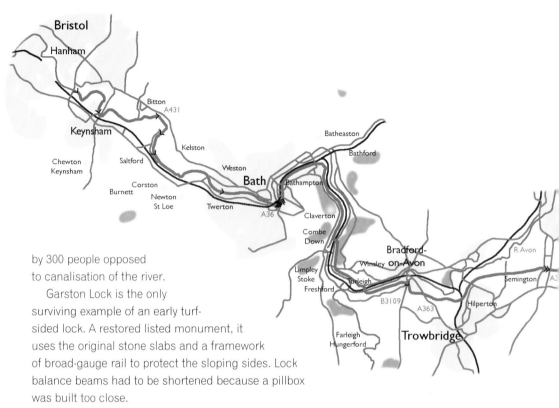

by 300 people opposed to canalisation of the river.

Garston Lock is the only surviving example of an early turf-sided lock. A restored listed monument, it uses the original stone slabs and a framework of broad-gauge rail to protect the sloping sides. Lock balance beams had to be shortened because a pillbox was built too close.

Sheffield Lock with its scalloped sides is set in an area of lawns. Scallops are in the original rail rib positions. Its takes its name from Sheffield Bottom, which gained fame when the local Stompers jazz group appeared on *Opportunity Knocks*, or might have done if Hughie Green had not announced them as the rather more intriguing Sheffield Bottomstompers. These days Hardy might have chosen a different name than Gaymead for Theale, in the opposite direction. It has Georgian terraces and a church built between 1820 and 1832 by EW Garbett in the style of Salisbury Cathedral, although not so obviously as the one at Bishops Cannings, further along the canal.

The swing bridge at Tyle Mill Lock is typical of those along this canal. These included some of the first applications anywhere of ball bearings. This one is now electrically operated. Tyle Mill itself was a flour mill, burnt down in 1914, rebuilt as a sawmill that was used until 1936 and is now a private house, neighbours including Kate Bush. The road leads out to the A4 and the Spring Inn, still known to many as the Three Kings Jacks Booth public house, partly named after a prize boxer and his booth.

A railway disaster that killed seven people was caused by someone deliberately stopping his car on a level crossing in the path of a high speed train at Ufton Nervet in 2004. On the next corner, at Lower Padworth, the canal, railway and A4 all come together for the first time. Had Brunel started his railway even five years sooner it might be that all our railways would

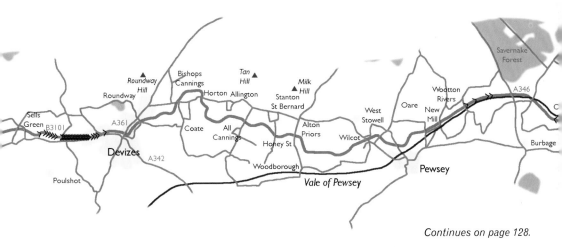

Continues on page 128.

now be broad gauge but standard gauge was already too well established elsewhere and he had to concede defeat. One weekend in 1847 the entire line from Paddington to Penzance was changed from broad to standard gauge with trains running on the new gauge on the Monday morning. It would be interesting to see how long Brunel would have taken to electrify the West Coast Main Line.

By Padworth Lock, the canalman's cottage has been converted into a Kennet & Avon Canal Trust visitor centre with a historical and wildlife walking trail.

In 1894, plans were announced to flood the valley between Aldermaston and Newbury as a reservoir for London. The plan was shelved after widespread opposition. Thames Water have been proposing more recently a site across the line of the Wilts & Berks Canal near Abingdon.

At the end of this straight the canal rejoins the river. To the left is Aldermaston, home of the Surridge cricket bat. On the other side of it is the Atomic Weapons Establishment, scene of the Ban

the Bomb marches in the 1950s. Between the village and research centre, Blue Circle built the new headquarters for their cement company, then had to vacate them as radioactive water seeped beneath them and towards the village. The village still uses the old tradition of a candle auction of church lands every three years, where bids are placed by sticking pins into a candle which then burns down, the rents going to charity.

On the next river section a low swing bridge has been replaced by the high-level Wickham Knights footbridge on interesting brick abutments. Near here was an angling notice banning just about every form of pleasure on the navigation, the word 'boats' subsequently being painted out from the banned list.

Woolhampton has a recent swing bridge although its fast current makes it a difficult area for powered craft. The Rowbarge Inn has been much enlarged since the 1970s when the landlord was Larry Naismith, known for his roles in Ealing Comedies although the picture he displayed in the bar was a rather severe portrait

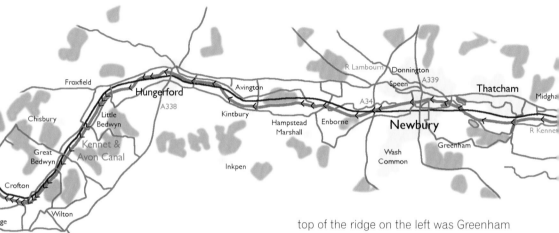

of himself in German military uniform from his leading role in *Where Eagles Dare*.

The station was renamed as Midgham to avoid confusion for passengers who really wanted to go to Wolverhampton, train announcers having their own version of English that did little to help passengers find the correct platform.

Woolhampton Lock was previously turf-sided but was restored with a brick chamber. Turf-sided locks were cheaper to build but used more water, something in short supply, especially further up the canal. This one is just above the point at which the river is left.

Thatcham probably takes its name from the marsh and reedbeds that provided material for thatching. It is one of the oldest Berkshire villages, with traces of a Mesolithic settlement and extensive working of flint, bone and antlers. It has had markets from the time of Henry I but suffered badly from the Black Death. There is a 14th century chapel near the Victorian church.

A straight avenue of reeds, occupied by reed buntings, leads to Widmead Lock. On top of the ridge on the left was Greenham Common airfield, where the Americans had Cruise missiles stationed from the 1970s. Their departure probably followed after the adverse publicity resulting from the women's peace camp at the gate. The airfield itself, which hosted an airshow during that period, has now gone too, an industrial estate spreading across the former site, but the odd dilapidated protest caravan is still occupied at the gate.

The railway finally crosses on a new bridge. The old pillars remain in place but don't quite meet the new bridge. Brunel would probably have loved this, exactly his sense of humour, the line he took when he was required to shore his bridge over the Thames at Maidenhead. Beyond the bridge there is a Nature Discovery Centre on the right before the River Kennet joins, having picked up the River Lambourn, as Bulls Lock is approached.

Beyond Ham Lock, adjacent to another recent industrial estate, is Whitehouse roving bridge where a breach occurred in 2000 near where contractors were digging footings for a new bridge. It resulted in a narrowboat going aground and being pinned across the breach.

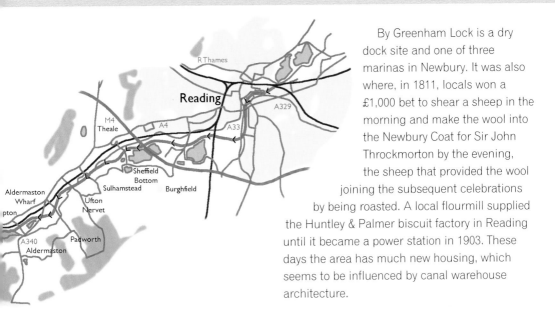

By Greenham Lock is a dry dock site and one of three marinas in Newbury. It was also where, in 1811, locals won a £1,000 bet to shear a sheep in the morning and make the wool into the Newbury Coat for Sir John Throckmorton by the evening, the sheep that provided the wool joining the subsequent celebrations by being roasted. A local flourmill supplied the Huntley & Palmer biscuit factory in Reading until it became a power station in 1903. These days the area has much new housing, which seems to be influenced by canal warehouse architecture.

Garston Lock, a turf-sided lock.

The A339 was formerly the A34, a traffic light-controlled roundabout creating a notorious bottleneck that typically took an hour to pass on what was the main north–south route in central southern England. Beyond the bridge is Newbury Wharf, or was until it was filled in during the 1930s to become a bus waiting area, then a car park, but may yet be partially restored. A crane imported from inside a railway building stands prominently in front of the Kennet & Avon Canal Trust's sales and display building in former canal stables. These may have been built from stone from the original castle erected in the early 12th century by the Earl of Perche, stormed in 1152 by King Stephen after a two month siege and later seized by King John.

An 18th century granary with an interesting balcony and the Jacobean Cloth Hall are used to house the Newbury District Museum. This features archaeology, ballooning, cameras, old costumes, weaving, butterflies and moths, a 2m elephant tusk from the Kennet and prehistoric and Saxon material and artefacts from the Battles of Newbury. Sean Keaney was born here in 1990 with 12 teeth, which had to be extracted so that he could feed.

Newbury was successively a Roman, Saxon, Norman and Tudor cloth centre. Most prominent was Jack of Newbury, hero of Deloney's story of the same name. He was apprenticed to a cloth maker, married the widow of his boss when he died, inherited the business when she died and employed up to 1,000 local people (150 of whom he took to fight for Henry VIII at Flodden) but declined a knighthood in order to remain on equal terms with his employees. He was subsequently visited by Henry VIII, Catherine of Aragon and Cardinal Wolsey.

In 1643 Cromwell's soldiers caught a witch sailing down the Kennet on a plank. She caught their bullets in her hands but they killed her by the usual method of slashing her forehead and shooting her below the ear. Windsurfers beware.

The Waterside Youth & Community Centre is beyond Parkway bridge.

DH Lawrence and his wife were resident here in 1918 after being evicted from Cornwall. Hardy was not at his most inspired when calling Newbury Kennetbridge but he did pick the most important feature, the balustraded stone 6.1m elliptical single-arch bridge of 1770 by James Clarke. This carried the A34 until 1965, and identified the town in *Jude the Obscure*. This section of river past the back of the Old Waggon & Horses is fast and narrow. Boatmen could be fined for allowing horses to haul across the road, there being no towpath under the bridge so that a long line had to be let down with a float attached. The authorities are still opposed to use of the road, which now has a very low speed limit and may only be used after 5pm. Newbury was the first town to have Belisha beacons. The towpath wall and bridge are well gouged with rope grooves. Next to the bridge is the Griffin butcher's shop selling second-to-none pork pies. Other significant buildings are the 19th century Italianate Corn Exchange and the Perpendicular church of St Nicholas of about 1500, with its unusual 17th century pulpit.

The Lock, Stock & Barrel is a reference to Newbury Lock, in front of which is the channel of the River Kennet. Newbury Lock was the first lock to be built on the man-made canal. Marked out alongside is the plan of the lock cottage that was occupied until 1958. This was leased by canal restoration pioneer John Gould. Another local canal business is the Kennet Horse Boat Company with motor trip boats here and horse-drawn boats at Kintbury.

The next two locks, Guyer's and Higgs, are named after the opposing commanders in the First Battle of Newbury. This was fought on Wash Common in September 1643, the Roundheads beating the Cavaliers in one of the bloodiest

The beautifully restored lock at Aldermaston Wharf.

battles of the Civil War. The Second Battle of Newbury took place 13 months later at Speen, when the Roundheads drove the Cavaliers out of Donnington Castle. The Cavaliers, however, returned the following week and relieved the castle, despite its anticlockwise staircases designed to put attacking right-handed swordsmen at a disadvantage.

The railway crosses back over Pickletimber Bridge and the canal approaches Benham Lock with a couple of pillboxes. In the summer, a blanket of buttercups extends in the direction of Enborne. The town is 2km from the River Enborne, which featured as an obstacle to the rabbits of *Watership Down*, written by Richard Adams, who was born in Newbury. To the north is Benham Park, laid out by Capability Brown, the house built 1772–1775.

Beyond Enborne Copse, Hamstead Park was built by the Earl of Craven, the house resembling Heidelberg Castle to please Elizabeth, Queen of Bohemia, to whom he was romantically attached. The scheme did not succeed and he died a bachelor aged 91. He was also a firefighting enthusiast, helped by the fact that his horse could smell fires and take him to them before the alarm was raised. Ironically his own house burnt down in 1718, 21 years after his death. Avenues run through the geometric gardens where the walls and six sets of gate piers remain. Mottes

West Mills Yard forms part of Newbury's waterfront.

beside the canal were part of the defence works of the 13th century Earl Marshall. Up the hill, the church is Norman with 14th and 18th century additions. The 19th century watermill by Hamstead Lock is probably on the site of one recorded in 1086. Newest of all, here and further along the canal, are gates where the towpath crosses the road. Crossing was not easy previously for those on foot because of the relatively blind road approaches but at least it was possible to escape on to the towpath. Now the user is trapped on the road with canoe, bike or pushchair while trying to undo Canal & River Trust gates.

Brick, flint and wrought-iron work at Hungerford.

Between Copse Lock and Dreweat's Lock, the ground rises from water meadows to a high ridge, a most attractive area of country, leading to Irish Hill, a corruption of *ebrige* or yew-covered ridge. Whiting Mills, closed early in the 20th century, used quarried chalk to make a fine white powder that was carried by canal to Bristol for use in paint manufacture. The Wilderness on the north side of the canal seems no more deserving of the title than many other sections of the canal.

In 1364 John of Gaunt gave the fishing and grazing rights to the town of Hungerford and donated his hunting horn, which is still in a local bank vault. Two years later he was given the manor, for which the token rent of a red Lancastrian rose is still paid to any monarch passing through the town. At Hocktide, the second Tuesday after Easter, a replica of the horn is blown to call 99 commoners to the Hocktide Court where two Tuttimen are elected. They carry Tutti poles decorated with flowers and ribbons and accompany the Orange Scrambler as he collects a penny a head from each commoner's house. He may kiss every woman in the town and they are each given an orange in return. Each new commoner is shod by having a nail driven into the shoe and everyone ends up in the Three Swans Hotel at lunchtime to drink Hocktide Punch to the memory of John of Gaunt.

Elizabeth I stayed in the Bear, on the A4. Charles I stopped there during the Civil War. William of Orange was staying there in 1688 when he received messengers from James II, rejected their proposals and continued his march on London. Old Mother Red Cap was a nurse said to have lived on the premises to the age of 120. Georgian coaching inns and 18–19th century houses are features of this normally quiet market town.

Decorative ironwork enhances a house by the A338 bridge. An open grassed area leads to a former granary and Hungerford Lock. Above this is a Gothic church of 1816 that replaces one that collapsed under the weight of a snowfall. Water meadows lead to Hungerford Marsh, an 11ha nature reserve that has recorded 120 species of bird.

Crofton pumping station, seen over the railway from the canal.

A metal footbridge over the canal continues over the railway with high flights of steps each side at Little Bedwyn, with its flint church of St Michael. The name has nothing to do with Arabs but comes from the Saxon *bedwine* or bindweed or from the Celtic *bedd gwyn*, white grave. The village had a Roman camp and Saxon mint. In 675 Escuin seized the throne from King Wulfhere in the Battle of Bedwyn. To the south is an 18th century farming village and there are 19th century patterned brick terraces to the north.

Beyond Little Bedwyn and Potter's Locks, Chisbury lies to the north-west with the 13th century thatched Chisbury Chapel, rescued from use as a farm building. Chisbury Hill Fort, referred to by Thomas Hardy as Batton Castle, is built on Bedwyn Dyke. This was part of Wansdyke, dug 138km between Inkpen and Portishead in the 6th century by Saxon King Caelwin as protection against the Angles from the Midlands.

By Bedwyn Church Lock is Bedwyn church, begun in 1092, with fine pierced embattlements around the top of the tower, built on the site of a 905 Saxon church. Close by is the Bedwyn Stone Museum with stone carvings, tombstones, statues, a dinosaur footprint and the Lloyd's stone masonry, which made the original inscription for the Bruce tunnel east portal.

Before Crofton Bottom Lock, the start of the Crofton Flight, the canal crosses the line of the Roman road from Cirencester to Winchester, heading south-east past a tumulus and the 1821 five-storey Wilton windmill, the only working mill left in Wiltshire. The other way leads through Savernake Forest. The 6km straight Grand Avenue of beeches with three other straight avenues crossing at its major intersection was cut by Capability Brown and runs parallel to a Roman road, perhaps acting as the inspiration. This Norman hunting forest is the only one still in private hands and roads through it are closed once a year to retain their private status.

The three hectares of Wilton Water are home to mallard, teal, pochard, tufted duck and moorhen. More importantly, it acts as a reservoir to supply the summit level, which Rennie constructed only 4km long. This was one of the shortest summit levels in the country when it was built and it is above the springline so it has no feeders. He built the Grade I Crofton pumping station to raise water 12m. Water difficulties mean that back pumping is required more than ever today. The Cornish beam engines are the oldest in the world still in steam and doing their original job. They are a Boulton & Watt of 1812 with 1.1m bore and 2.4m stroke and one by Harveys of Hale in 1845, their efficiencies measured at 1.8 and 2.4 per cent respectively.

Modern pumps do most of the work today but the beam engines did have to be used on their own for a couple of days in 2009 when the electric pumps failed.

Henry VIII courted Jane Seymour at Wolfhall, or Wulfhall, with the king taking over the house here while his prospective inlaws had to move out into a barn.

Tottenham House, set in a deer park to the north of the canal, has a monument to the restoration from insanity of George III, erected by the Marquis of Ailsbury. The current Marquis of Ailsbury is a direct descendent of Richard Sturmy, who is recorded in the *Domesday Book* as holding seven manors in Wiltshire.

Bruce Tunnel, 460m long, is named after Thomas Bruce, the Earl of Ailsbury, who

Lady's Bridge at the end of Wilcot Wide Water.

Avoncliff Aqueduct seen from the railway span.

the centre of the span, and hanger chains that become less vertical towards the centre.

A smart village cricket pitch fronts Wilcot. The village also has thatched cottages, the Golden Swan's having the steepest thatch in Wiltshire. A manor and vineyards originated in the 11th century and there is a school with a prominent bell as well as a 12th century Holy Cross church with a Norman chancel arch, mostly rebuilt in 1876 after a fire.

wanted a tunnel rather than the proposed cutting through his deer park. It is second only to Netherton for the size of bore of any tunnel remaining open and has chains along the sides for pulling boats through as legging would not have been possible.

Cadley and Brimslade Locks mark the start of the Wootton Rivers Flight. Brimslade Farm has 17th century tiled buildings with many chimneys. Rivers of any size are not in evidence, however, as the canal cuts through the Vale of Pewsey to Devizes with Lower Greensand and clay, surrounded by chalk hills, in a conspicuously dry valley. The canal also runs parallel with the Wansdyke earthwork to Devizes.

Wootton Rivers is a village of brick and half-timbered thatched cottages. The church, which was rebuilt in the 19th century, has a wooden bell tower and a clock made by Jack Spratt to commemorate the coronation of George V. It uses junk such as bedsteads, prams and bicycles and has the letters GLORYBETOGOD instead of numbers on the face and 24 different quarter-hour chimes.

From Stowell Park, with its 19th century house, the Stowell Park Suspension Bridge crosses the canal. The bridge is not spectacular in appearance but is one of only two James Dredge bridges remaining in England, both over this canal. The light structure has suspension catenary chains that reduce in thickness towards

Wilcot Wide Water and the neoclassical balustraded Lady's Bridge across one end were features built to appease estate owner Lady Susannah Wroughton. The bridge leads towards Swanborough Tump, where Alfred the Great is believed to have held a Parliament and written his will while being attacked by the Danes. The Bronze Age Ridgeway crosses somewhere in this vicinity, also passing the distinctive Picked Hill and Woodborough Hill with its strip lynchets, relics of Celtic and medieval cultivation. Cylindrical anti-tank blocks on the end of a bridge near Woodborough are of a kind seen at several canal crossing points.

Alton Priors, at the site of a Stone Age camp, has a Perpendicular church with a large 1590 box tomb, on which is an interesting Dutch brass plate. Neighbouring Alton Barnes, mostly owned by Oxford's New College, has an 18th century rectory and a heavily restored Anglo-Saxon church with everything in miniature.

Because of the lack of subsequent building, many more ancient sites have survived on the Downs than in most other parts of the country.

Dundas Aqueduct with its cycle track.

Dundas Wharf looking towards the end of the former Somerset Coal Canal, now a spur used for moorings.

Knap Hill has a 1.6ha neolithic earthwork enclosure intersected by many causeways. A 2,000m² rectangular earthwork on the east side was a protected Iron Age homestead. There are also neolithic tombs and Adam's Grave, a prehistoric barrow.

Weather-boarded buildings and brick cottages make up Honey Street, where there are also a warehouse and the ruins of a boatbuilding yard that launched canal craft and sailing trows sideways into the canal, the remains of one being *Harriett* among the Purton hulks. Up until 1948 there was still a floating bridge across the canal on the right of way leading to the Tan Hill fair, replacing an earlier ferry. Also important is the popular Barge Inn, which had a brewery, bakery and slaughterhouse and is now a rock-and-roll pub. After burning down in 1858, during which time the locals attempted to consume the cellar's contents, it was rebuilt in six months. It also has a room devoted to crop circles, with details posted as they become known. Typically, ten

circles are reported on successive days at the beginning of June and enthusiasts from around the world gather in this building to get the latest news. Activity tends to centre around Avebury with the canal being situated along the southern edge of the area of sightings.

There was a large 7th century Iron Age settlement with iron smelter and storage pits at All Cannings Cross. A Neolithic-style long barrow was built at All Cannings in 2014 to hold 300 urns of modern cremation ashes. The 14th century All Saints church has an 1867 High Victorian chancel. The canal banks are lined with blackberries and convolvulus.

The canal loops round Horton, part of a 38km² Crown estate. A high footbridge crosses the canal to Bishops Cannings, named so because it belonged to the bishops of Salisbury. St Mary's church is like a miniature Salisbury cathedral and has a spire rising to 41m, a cruciform plan, rich stone carvings and Early English traces from 1150.

The village has thatched and half-timbered cottages, including the 17th century thatched farmhouse of the Old Manor and the 17th century timber-framed Old School House. The tale of Moonrakers, locals claiming to be raking the moon from the surface of a piece of water when surprised by excisemen, comes from various places around the country but Bishops Cannings' pond is the most likely venue as it was on the route used to move Dutch gin from the Hampshire coast to Swindon by night. Wiltshiremen are still called Moonrakers.

Horton Bridge (which is not the nearest bridge to the centre of Horton) has a Bridge Inn. From here the canal turns west again past Coate, a source of puddle clay for lining the canal. This section of canal is below Roundway Down where, in 1643, during the Civil War, the Royalist cavalry made a surprise attack on the Roundheads, killing or capturing most of them.

The A361 makes its first crossing as the canal turns into a straight that runs past the hospital to Devizes Wharf, which is now mostly a car park. This is the site known to canoeists as the start of the Devizes to Westminster Race. The DW, Britain's best-known canoeing event, began in 1948 in answer to a £20 bet over whether the local Rover Scouts could travel to Westminster by kayak in 100 hours. They won the bet and the subsequent race has developed into an annual Easter activity that has been instrumental in encouraging restoration of the canal, Britain's major canal restoration project. The 1979 race record of 15 hours 34 minutes still stands. Along one side of the wharf is an 1810 granary with a long timbered balcony that houses the canal centre and canal society, while the Wharf Theatre is located in the warehouse.

Devizes has the largest marketplace in the west of England, chartered since 1141 and much influenced by its surroundings, taking its name from the Old French *devises* meaning boundary. The Wiltshire Heritage Museum has finds from prehistoric, Neolithic, Bronze and Iron Ages and Roman material with items from

barrows, Avebury and Stonehenge – the Bronze Age collection being one of Britain's best. Many houses are Georgian, which made this a suitable venue for filming *Far from the Madding Crowd*. The market cross bears the tale of Ruth Pierce, who was involved in a three-way dispute over part of the payment for a sack of wheat in 1753. She asked to fall down dead if she was guilty and duly died, the missing 3d being found in her hand.

The Devizes Flight is one of the wonders of the canal system. While Tardebigge has a greater number of locks, here there is a greater fall in a shorter distance, Britain's longest continuous flight. The 29 locks drop the canal 72m in 3.6km. At Dunkirk, the Black Horse had a major fire in 2010, which was extinguished with water from the canal. The A361 crosses over Prison Bridge, near the former prison, and the towpath again has a separate arch. The most dramatic section is the Caen Hill Flight, which runs straight down a spur of land at a 1:30 slope. Sixteen locks are closely spaced, each with a 64m x 41m x 1.6m deep widening between locks. The flight was completed in 1810, the last part of the canal to be finished and the last part to be restored.

While Seend Cleeve is now a quiet agricultural village, this was not the case in Victorian times when there was an ironworks halfway down the Seend Flight of locks. Three hundred tonnes a week of iron was mined from higher ground and transported by canal to south Wales, later being processed here in two blast furnaces until 1889. Patrons at the popular Barge Inn on a summer's day would find it hard to believe the change in the scenery.

After an aqueduct for the new route of the A350, Barratt's and Buckley's Locks come in quick succession at Semington. An attractive lock keeper's cottage remains, unlike the Wilts & Berks Canal, of which the bricked-off end is located on the right before the bridge carrying the old line of the A350. This canal, which ran to the Thames at Abingdon, is now the country's major restoration project. A line along the River Avon will be used in part. It is hoped that the whole canal will be ready by 2025.

The Semington Brook passes under Semington Aqueduct and the brook and the canal both pass the village of Semington with the first of the Cotswold stone houses that are to replace brick from here. Some are 18th century and there is a stone church with a bellcote and an adjacent village school in similar style.

The weaving town of Trowbridge, named from its wooden bridge, the Old English *treow* meaning tree, with its 18th century houses

Rennie's unique waterwheel-powered pumping station feeding the canal up the hill to the left.

and lock-up, has become the administrative capital of Wiltshire. It was served by Hilperton Wharf, which has now been converted into a mini Docklands housing development, reached under the re-sited Parson's Bridge. The futuristic Alvechurch Boat Centres' marina has been built on the other side of the canal. Wyke House, with its Jacobean-style towers, is a replica of 1865. From 1814 to 1832, poet George Crabbe was the local rector.

Bradford-on-Avon is a market town in a steep, wooded valley, developed with the help of Dutch weavers, there having been 30 weaving factories in 1800. One of the most remarkable buildings is St Laurence's church, founded in 705. This is the most complete Saxon church in England, which

was rediscovered in 1856 by a vicar on a hillside who noticed the cross in the roof patterns. It was largely intact, despite being used as a school and a cottage and having previously been used as a slaughterhouse. More conventional is the 12th century Holy Trinity church with additions, wall paintings and fine 18th century monuments. The 13th century Oratory Bridge over the river, rebuilt in the 17th century, has a pilgrim's chapel, which was used as a lock-up for a while, John Wesley being an inmate for a night in 1757. It is topped by a copper-coloured gudgeon weathervane. Other significant buildings include the Victorian town hall and a Gothic revival factory.

The pedestrian route past Bradford-on-Avon Lock, one of the busiest in the country, is long and difficult; it involves crossing the B3109 bridge with blind approaches for vehicles, passing the Canal Tavern, which was used to stable horses, and relaunching past tables of customers who sit outside in the summer taking morning or afternoon tea. From here to Claverton the canal corridor is noticeably well used for recreation. The line from here to Bathampton offers some of the best English canal scenery.

On the right, beside the canal, is the Grade II Barton Tithe Barn of about 1341, which formerly belonged to Shaftesbury Abbey. One of the largest in England, it is 51m long, has two porches with massive doors and is beamed with a cruck stone-tiled roof, one of the largest in Europe, covering 930m^2.

Below Westwood, with its 15th century stone manor, noted for its late Gothic and Jacobean windows, is Avoncliff, where there are 18th century weavers' cottages in a square, the Cross Guns 15th century coaching inn, the chimney of a former flock mill and one of two mills, retaining its enormous wheel.

The canal now switches to the other less steep side of the valley for a while by crossing the Avoncliff Aqueduct. A neoclassical structure in Bath stone with balustrades at the ends, the

Cleveland House, the canal company headquarters.

three-arched aqueduct is higher than its near neighbour and longer at 159m. With subsidence in the centre, it has an irregular line to the eye from on top, especially as it kicks sideways over the railway. Leakage meant that an icicle remover had to be employed in winter to clear the section over the railway each morning where the icicles posed a danger to trains. A tramway from quarries in Becky Addy Wood ran across the aqueduct as a feeder for the canal.

The canal passes from Wiltshire into Bath & NE Somerset, arriving at Dundas Wharf with its toll house, small warehouse and cast-iron crane, which worked until the First World War. At the southern end of the wharf is the narrow entrance to the Somerset Coal Canal, now only a 500m spur serving as moorings although it was noted for its inclined planes when operational. At the northern end of the wharf is Dundas Horse Bridge, a Grade II roving bridge partly from 1794.

Above Claverton church is Claverton Manor of 1820 in Greek revival style by Sir Jeffrey

Wyatville, once owned by a Shaker family. It was the location for Churchill's first political speech in 1897, when he was 23. It is Europe's only museum of 17–19th century American domestic life, being partly 17th century itself.

Bathampton has a 19th century church but the haunted George Inn is 14th century, formerly part of a priory, and had to have its door moved to the side as the canal passed so close. It was the location of the last duel fought in England. A huge collection of gnomes in a garden will not be to the taste of everyone.

Large gardens with an air of affluence slope down to the canal. This part of the canal, as it heads south-west, gives the best views over the Iron Age hill fort on Solsbury Hill and the dazzling white stone housing of the city. Below, on the river, is the Victorian Bath Boating Station, which offers skiff, punt and canoe hire. Higher up, space is more cramped and part of the canal had to be moved to the left in 1839 to accommodate the GWR. Sydney Gardens were created in 1794 and hid the canal when it arrived in this fashionable part of Bath, further concealment taking place in the meandering hedges of the maze, which was to go on to inspire the one at Glendurgan.

No. 2 tunnel is 50m long with an Adamesque portal and carries the A36, McAdam's 1830 turnpike, over the top. Two cast-iron footbridges were imported from Coalbrookdale in 1800 and the canal is noted for its Chinese-style bridge. The final tunnel is 54m long and is topped by the Georgian Cleveland House, a stone building of 1820 by the Duke of Cleveland. It was the original headquarters of the Kennet & Avon Canal Company and has a trap-door from the cellar. The adjacent Macdonald Bath Spa Hotel distils its own whisky. Nearby is the Holburne Museum of Art in the former Palladian Sydney Hotel, which sets the tone well for Bath.

Pulteney Weir upstream of the canal junction with the Avon.

From Sydney Wharf, the Widcombe Flight continues right down to the Avon, starting with an interesting lock keeper's cottage and cast-iron footbridge of about 1815 by the Bath Top Lock. Halfway down the flight is the ornamental chimney with 2° lean for a pump engine house and there was another at the bottom lock. Side ponds were dug into the hillside for the locks. After Pulteney and Abbey View Locks, the A3062 crosses and then Wash House Lock has a footbridge of about 1815. Roadworks have resulted in Bridge and Chapel Locks being combined into Bath Deep Lock, the second-

deepest inland lock in Britain at 5.7m. The route on the left goes down a flight of stairs into a tunnel with a piece of railing at the bottom, which is unhelpful to anyone portaging. The alternative route is to brave the traffic. Beside Widcombe Lock, the Thimble Mill pumphouse has become a restaurant.

The canal feeds out on to the River Avon, already quite broad, heavily piled and potentially muddy. The River Avon and the south Wales rivers may have originally been dip slope headwaters of the River Kennet, accounting for the canal's route. There were legal powers

to make the river navigable for large craft from 1619, although the work was not completed until 1723. There were plans to carry the canal on downstream but concerns that it might interfere with Bath's hot spring water meant the idea was not pursued. In fact, the river remained profitable with traffic between Bristol and Bath until 1967 as the canal itself declined. The river has a 6.4km/h speed limit and downstream craft have priority.

The first bridge over the river is the lattice Widcombe Footbridge, a footbridge also known as Halfpenny Bridge because of its toll. In 1877,

The Roman Great Bath with Bath Abbey behind.

during the Bath & West Agricultural Show, it collapsed, resulting in the deaths of 10 people and injury to 40–50 others as they queued on the bridge to pay the toll following the arrival of a train in the station opposite. The station was built by Brunel in a Jacobean-country-house-style, with curved gables and a castellated entrance to the viaduct at the west end of the station. It originally had a hammerbeam roof. His GWR Paddington to Temple Meads railway crosses the river on St James' Bridge and then runs on to the 550m St James' Viaduct, which has 73 arches, decorated with castellated octagonal turrets, typical Brunel ostentation. Adjoining the station is the Graze restaurant with the Bath Ales microbrewery.

The iron arched Cleveland Bridge dates from 1827. The major bridge interest is Robert Adam's Pulteney Bridge of 1773, based on Palladio's rejected design for the Rialto Bridge in Venice, the only British bridge with buildings on both sides. Below it is Pulteney Weir, with a mill at each end, rebuilt as a unique parabolic stepped structure in 1965 but now dangerous because of erosion under the bottom step and due for a further rebuild. The weir can be reached upstream from the canal.

Bath has been a World Heritage City since 1987. In 863 BC Bladud, the son of King Lud Hudibras, the father of King Lear, caught leprosy and worked as a swineherd but the pigs caught it too. The pigs went into a warm black swamp and were cured so Bladud did likewise. The carved acorns around the city recall the pigs' favourite food. The springs were suggested as the gateway to the underworld. Bladud built Kaerbadum, now Bath, constructed the hot baths, chose Minerva and encouraged necromancy. This was the beginning of the taking of the waters for their curative properties. To the Romans it was Aquae Sulis after the local god. Britain's only hot springs, 2.9l/s emerging at 46°C, were used to supply the Roman baths that were used from 65 to 410 and from which the city takes its name. A curse thrown into the King's Spring among 12,000 coins includes the earliest reference in England to Christians. When rediscovered in 1775, the spa produced the greatest collection of Roman finds in Britain: the temple of Sulis Minerva; mosaics; statues; jewellery; pottery; tombs; saunas; cold plunges; and the Great Bath, 22m x 9.1m x 1.5m deep, which still uses the Roman plumbing and its 8.5t of Mendip lead lining, this being one of the world's finest ancient thermal spas. The Gorgon's head from the temple, actually a male, has been copied as the centrepiece of the maze in the Parade Gardens.

The See of Wells moved here in 1088 and Bath Abbey, actually a priory although once a cathedral and wool centre, as recalled by Chaucer's Wife of Bath in *The Canterbury Tales*, was begun in 1499 on the site of a church of circa 680, where Edgar was crowned in 973 as the first king of England. The church was gutted by fire in 1137. The 1499 design was given to Bishop Oliver King in a dream, resulting in the stone carvings of angels climbing Jacob's ladder to and from heaven on the west front. In a piazza, the Perpendicular Gothic building has fine fan vaulting, restored 18th century cellars with an exhibition of 1,600 years of abbey history, a 49m tower, a heavily carved oak door and a wrought-iron screen. It is called the Lantern of the West because of the 52 windows, especially the brilliant 17th century 76m^2 window.

The oldest building in Bath is Sally Lunn's House of 1482 although it was not until 1680 that she baked her Hugenot brioche buns here. They are still baked to the same recipe. The house has a museum of Roman, Saxon and medieval building and clay pipes in the cellar. Bath buns are sugar-topped, spiced, currant buns, another Bath speciality. Bath Oliver biscuits were invented in the 18th century by Dr William Oliver as part of an obesity diet.

The city was the first British tourist resort and considered fashionable after a visit by Queen Anne. It became established as a spa in the 18th century by dandy Beau Nash, who was known as the King of Bath for 50 years, banning the wearing of swords and other anti-social behaviour. The pump room was opened in the style of an orangery in 1706 and rebuilt in its current form in 1789–1799 by Thomas Baldwin.

Exhibits include a 3m longcase clock of 1709 and two sedan chairs, cheaper to operate than the three-wheeled Bath chair, invented locally by Arthur Dawson and able to take the user indoors and up stairs if disabled. A Fashion Museum is located in the Assembly Rooms. The water flows at 13l/s with a high content of calcium sulphate and sodium chloride but low in metals except iron. Many of those treated with the waters were suffering from lead poisoning, especially from exposure to it in various industries and from its use to sweeten and preserve alcohol. Thermae Bath Spa, opened in 2003, is in a glass and stone rooftop building with indoor and outdoor thermal pools, a whirlpool, neck massage jets, airbeds, steam rooms and massage rooms. It is the only UK bathing site that uses natural thermal waters.

Despite having been heavily bombed in the Second World War, the city offers a wonderful Georgian townscape, much of it designed by John Wood the Younger. The Guildhall of 1766–1775, by Thomas Baldwin, with its Reynolds portraits, is claimed to be the finest room in Bath. In the autumn it is one of the hosts of the Mozartfest. The city also hosts a spring Bath International Music Festival.

The market is on the site of a medieval slaughterhouse with a 12-sided domed interior. The Museum of Bath at Work tells the story of Bath's industrial, commercial and social history and much more besides. It features exact reconstructions of factories and workshops.

Dredge's suspension arrangement in the Victoria Bridge.

Kelston Brass Mill and the weir.

There is a book museum, while the Victoria Art Gallery features 18–20th century British art, notably works from Turner, Gainsborough, Sickert and Whistler, and also prints, drawings, ceramics, glass and watches.

Local points of interest include the Forum, the Impossible Microworld of Willard Wigan and a sculpture of huge nails. Even Sainsbury's store has a Georgian frontage with Ionic columns and octagonal columns holding wrought-iron arches, this being a leading example of a lesser Victorian station, the Midland Railway's Green Park terminus of 1870.

The William Herschel Museum is in the Georgian furnished home of astronomers and musicians William and Caroline Herschel, and is where Uranus was discovered in 1781. Stone buildings are all around but none can match the 200m Royal Crescent of 1767–1774 by John Wood the Younger. The 1796 home of the Duke of York at the end is now a Georgian museum and it is said that the ghostly footsteps of an unpaid servant who left two of the duke's illegitimate children to starve can be heard there. It is set in the 23ha Royal Victoria Park, a Park of Special Historic Interest with lakes, an aviary, a botanical garden, bowls, tennis, a children's play area and the Dell, with ghostly sounds from the former duels for which it was used. The Georgian Garden has plants and a layout from 1760. Bath Festival Maze of 1984 has a 4.5m diameter central mosaic incorporating the Gorgon's head and seven small mazes, which use 72,000 pieces of Italian tesserae.

Dutch Island takes its name from a former miller. Weston Cut goes right past the Dolphin Inn to Weston Lock, also known as Newton St Loe Lock. As with the following locks, smaller craft need to take out up-ladders. Unlike the following locks, this one bypasses a weir with large lifting radial gates.

The Chequers Inn, The Boathouse bar and Old Lock & Weir Ale House overlook the weir stream past Hanham Lock.

The ridge rising to the north of Keynsham Hams.

The former Newbridge railway bridge carries the Avon Walkway and cycleway over although there are plans to restore the Avon Valley Railway along this line to the outskirts of Bath. This is the end of the city with Bath Marina, the Boathouse public house and the A4 crossing between them on the graceful 26m Bath stone-arch New Bridge of 1740, which continues as a viaduct of stone on the south-west side of the river.

A dormitory suburb, Saltford has some interesting and attractive buildings near the river. Saltford Brass Mill, formerly a fulling mill, was water-powered, the last of its kind. It closed in 1925, the most complete annealing and brass-rolling mill in the country. Brass mills in this valley made their money out of trinkets for the slave trade. Handel is said to have written the *Hallelujah Chorus* after hearing the hammers. By the church, Saltford Manor is one of the oldest inhabited houses in England, its 17th century facade hiding what is a mostly Norman house.

The weir is about 1m high and slopes, more gently to the right of the chute in the centre. Kelston Brass Mill and annealing ovens add interesting architecture and the adjacent workers' cottages are some of the oldest known in existence.

The Avon Valley Country Park proves popular, with mini steam-train rides on the 127mm gauge Strawberry Line, quad bikes, adventure play area, barbecue area, pets corner and a falconry that can be seen from the river.

Keynsham was known to the nation's youth in the post-war years as it was spelled out several times per night on Radio Luxembourg in support of Horace Batchelor's football pools-beating scheme. It would seem that anyone with a surefire system should be using it for his own bets rather spending money to advertise it to the world in general. It inspired the title of the Bonzo Dog Doo-Dah Band's *Keynsham* album.

The river makes a sharp turn in front of Hanham Court and enters a dark Pennant sandstone gorge. Some of the craft moored along this reach are of substantial size. The banks are just as popular because of the Chequers Inn, The Boathouse bar and Old Lock & Weir Ale House. Hanham Toll Office was built at Hanham Lock, the start of the Port of Bristol.

The weir can be shot in suitable craft if conditions permit. This was the previous tidal limit but high tides can still reach here or even Keynsham, despite construction of a weir at Netham. Even powered craft can pass over it on high tides, when it drowns.

18 Grand Western Canal

The Grand Western Canal was part of a scheme to link the Bristol Channel with the English Channel. The Taunton to Topsham section was authorised in 1796 and John Rennie built the Greenham to Tiverton section between 1810 and 1814 as a single-level, wide beam barge canal, following the contours. The section through to Taunton was opened in 1838 and was an engineering marvel, having one inclined plane and seven vertical lifts, the only place where these really worked successfully. The lifts required trains of box like tub boats carrying 4–8t each. The lower section was disused by 1867 although signs of the lifts are still visible at Nynehead.

The lower section remained in use until 1924, when there was major leakage, being officially abandoned in 1962. In 1970 it was taken over by Devon County Council (in whose county the whole wide-beam section of the canal lies) and restored as a country park, including returning water to a dry section. Someone who made use of the canal was marathon runner Jo Purvey, who did her Olympic training along the towpath. Because the waterway is not connected to the rest of the canal network, boats have to be brought overland. It is therefore a nice quiet and clean place.

The canal is featured strongly in the Tiverton Museum, together with a 200-year-old forge, the first lace-making machine and railway equipment.

The landscaped terminus of the Grand Western Canal.

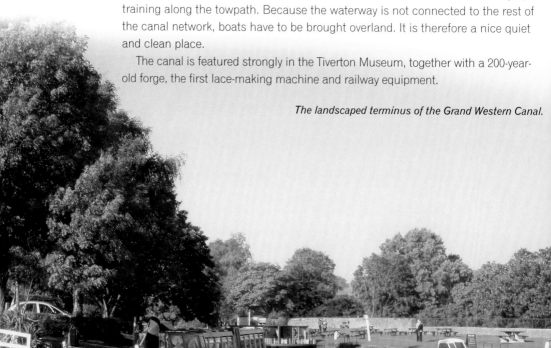

Autumn dawn mist east of Tiverton.

The missing link in the coast-to-coast canal was the descent from the canal basin in Tiverton (Hardy's Tivworthy) to the River Exe and then its canalisation to Exeter. The scale of the problem is easily gauged from the Tiverton basin, which looks down on the roofs of the town and commands fine views across to the hills northwards.

From the road at the northern side the complex resembles less a canal basin and more a railway terminus with its heavy, arched walls, although the two massive buttresses are actually old limekilns that were supplied with limestone from Burlescombe quarries by barge; a far cry from the wool and lace on which the town based its early prosperity from Edward III onwards.

These days the canal basin has neatly mown lawns laid out with picnic tables, planted trees, Ducks Ditty floating cafe bar, car park, toilets, 2012 visitor centre and the Sea Cadets' TS

Hermes, all overlooking an immaculate bowling green. The south bank houses the stables, offices and tack room of the Grand Western Horseboat Company, which takes horse-drawn barges nearly to the aqueduct.

This is one of those rare canals where there is no sign of modern industry; it mostly passes

Distance
18km from Tiverton to Greenham
Highlights
Tiverton Basin, horsedrawn trip boats and kilns
Navigation Authority
Devon County Council
Canal Society
Grand Western Canal Trust
www.grandwesterncanaltrust.org.uk
OS 1:50,000 Sheet
181 Minehead & Brendon Hills

A wooded section to the east of Tiverton.

through open and scenic countryside and even has attractive surroundings in the built-up sections, particularly the first kilometre leaving Tiverton where gardens leading down to the canal are obviously kept with pride. Greater reedmace, water lilies and arrowhead grow in the water. Oak trees provide sheltered sections to the canal at intervals. There are many ducks on the canal and swans can be a nuisance.

The bridges are all named, the first being William Authers, which is a laminated wood arch. Views are often good from the water along the canal but almost always only on the north side. As soon as the houses are left behind, it is possible to see across the Lowman valley to the hills beyond.

Dredging in 2003 turned up the remains of a Canberra bomber that had crashed into the canal in 1961 on a training flight from Germany, during which it suffered engine failure at low level. The two crew were killed after turning it away from Tiverton to more open ground.

Crown Hill aqueduct carried the canal over the railway connecting Tiverton with Tiverton Junction, used by the Tivvy Bumper. Unusually, the canal has outlived the railway.

After the bridge carrying what is now a minor road, the arch of which stands directly in front of an oak tree, the canal doubles back to the left. It then goes round three sides of a square to reach Halberton, its contour route taking it past a golf course and steep hillsides to the north of the village. The opening Dudley Weatherley Jubilee Bridge allows foot access across the canal. An embankment crossing a valley that divides three ways gives amazing views towards Noble Hindrance but suffered a major collapse in 2012.

Weed in the canal is water fern, a brown carpet that collects on the bows and can be a problem, despite its beautiful appearance when examined closely.

A pair of horse chestnut trees in the towpath hedgerow outside Sampford Peverell are in memory of a couple of two-year-old children. The good-quality towpath is well used by runners and commuting cyclists.

Weed removal is carried out from a weed-cutting boat. Some of the growth is yellow iris, enhancing the canal in the spring, with a backing of lovely green weeping willow trees. The Merriemeade and the Globe Inn are near the canal in the village.

A slipway has been added by the A361 bridge. The M5 passes within 700m of the canal but is not noticed, the only obvious heavy traffic being over the bridge carrying a minor road. Aircraft noise is another matter with naval jets on low-level flights.

Indicative of quieter days is the distinctive old chapel at Ayshford, which stands at the start of the first of the two long straight reaches on the canal. The Exeter to London railway sweeps past the second of the two bends connecting the straights, overlooked by the village of Burlescombe with its prominent church tower. In turn, it overlooks the meadows on the inside of the bend. Instead of flowing northwards to the Tone, excess water weiring over from the canal is carried underneath the channel and flows southwards as Spratford Stream to join the River Culm.

A bridge that once carried a railway spur to the quarries at Westleigh is now unused. Surprisingly, the bridge's girders are supported on a wooden crossbeam, which is now showing the passage of time.

Powerlines cross over and back and an electricity substation is located in the meadows.

Ivy-covered Whipcott Bridge.

After a couple of leads off to the left, the water becomes shallow, weed-free and absolutely clean. On the left bank are Wayman limekilns, set in a cutting. For the first time Canadian pondweed becomes the plant of the canal.

Without warning, the canal plunges into the short Wayman Tunnel at Beacon Hill.

Advance notice of the end of the canal arrives with a short area of reedmace right across the water, which comes to an abrupt finish just beyond in Lowdwells Lock, the first on the canal. An overgrown hollow leads towards the nearby River Tone and the Somerset border, leaving a major restoration project for the future although a portage of 300m will enable the canoeist to continue to the Bridgwater & Taunton Canal.

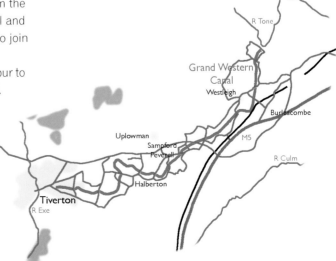

Bridgwater & Taunton Canal

The Bridgwater & Taunton Canal links the River Tone at Taunton with the tidal River Parrett at Bridgwater. It was once part of the Parrett, Tone, Chard and Grand Western navigations, from which there were plans for a link to Topsham, connecting the north and south coasts. The line from Taunton to Huntworth was opened in 1827. Bridgwater Dock was added after the completion of the canal in 1841 by Maddocks. Most revenue came from through traffic going to the Grand Western and Chard Canals. In 1866 it was sold to the Bristol & Exeter Railway Company, whose line closely follows the canal between the two towns.

Commercial traffic ceased in 1907 and the locks fell into disrepair. Somerset County Council purchased the dock for recreational purposes. The Somerset Inland Waterways Society was formed in 1965 to undertake the restoration, two years after British Waterways took over the canal, which is wide, open and clean.

Firepool Lock in Taunton, the first on the canal.

The Norman Taunton Castle, its Great Hall built in 1245, was defended by the Parliamentarians against 10,000 Royalists. Used for the law courts, it was where Judge Jeffreys condemned 500 men in the Bloody Assizes after the Monmouth Rebellion. It now has the Somerset County Museum including the Somerset Light Infantry with an American flag captured by them in 1813.

Downstream is the former wool and cider town of Taunton, Hardy's Toneborough. The canal leaves the River Tone at Firepool Lock, with a mere 500mm drop. It is tucked in a corner beneath tall trees and the first of the red-brick arch bridges, overlooked by the noisy cattle market and the railway goods yard and water tower that was fed from the canal.

The canal breaks out into open country. Until Bridgwater, the canal passes through nothing larger than the occasional agricultural hamlet and meets few main roads. It is an area of attractive rolling Somerset countryside, neither Sedgemoor and the Somerset Levels, which it skirts, nor the Quantock Hills, around which it arcs and which can be seen on the left at this point and on numerous later occasions. Banks are usually low, allowing extensive views.

Beyond the M5 is the site of railway water troughs and then a large brick mill on the banks of the River Tone is opposite the first houses of Creech St Michael. On the downstream side of the mill is an aqueduct across the river, one of the few remaining signs of the Chard Canal, which joined the Bridgwater & Taunton here. Operational only from 1842 to 1868, it was something of an engineering marvel with no less than four inclined planes along its length.

This point is also marked by the first of the pillboxes that line the canal. During the Second World War the War Office requested that the canal be turned into a defence line. So, in 1940, the fortifications were added and swing bridges were fixed and strengthened to carry

Distance
23km from the River Tone to the River Parrett
Highlights
Taunton Castle, including the Somerset museum
Someset Spacewalk
Blake Museum
Albert Street Cutting
Bridgwater Docks
Navigation Authority
Canal & River Trust
OS 1:50,000 Sheets
182 Weston-super-Mare
193 Taunton & Lyme Regis

War Department vehicles. These have had to be raised to allow canal boats to pass underneath.

The Taunton area is known for its cider. Apple orchards dot the landscape. The land rolls gently, here a hillock with a wind pump on top, there a dip below canal level, making the scenery more interesting, bringing the Quantocks back into view or giving glimpses of the unusual church at North Curry, apparently in Regency style.

The canal now moves away from the River Tone as it becomes tidal. Another divergence comes on the railway, which has been close for most of the journey but only visible when trains are passing. Even then, they are often heard but not seen. A strange asymmetric railway bridge of 1932 is the site of an oblique junction of major importance. While the Exeter to Bristol line continues to follow the canal, the shortened London route cuts left and then crosses under the more northerly line.

As well as the locks at each end, there are four locks – Higher, Maunsell, King's and Standards – grouped over 3km midway along the canal. These have totally unique features in the form of spherical counterweights on the lock gates; their concrete balance beams are also rare. A large orange and yellow sun sphere,

Creeper shrouds the bridge at Creech St Michael.

puzzling for passing motorists, is the centre of the Somerset Space Walk, a model of the solar system laid out both ways along the canal with planets and distances to scale. The upper gates at Maunsell Lock allow water to weir over and this can be disconcerting as the drop is substantial.

At North Newton the canal turns sharply in front of a meadow with a very ancient and picturesque church behind. This is not the village's only gem. In 1693 the Alfred Jewel was found in North Newton and this is now the oldest surviving Crown Jewel.

From Standards Lock it is possible to look right across the Somerset Levels and just see that most mystical of all the hills in Britain, Glastonbury Tor.

At Huntworth, present-day activity returns with the Boat & Anchor Inn and by the orderliness of the pairs of rectangular pillars that support the 900m long M5 viaduct that crosses the canal, railway and newly arrived River Parrett in one long glide. It is a surreal experience to sit under the viaduct at the canal end and look down that long line of columns as the invisible traffic roars overhead.

The canal enters the south end of Bridgwater and passes a brightly painted YMCA centre. A weir on the right is unprotected although signs warn of

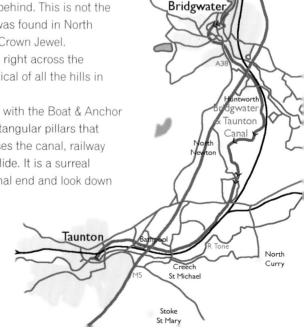

reduced headroom on a bridge that is no lower than many others on the canal.

Between a water intake and a high fence round what could be a tennis court, another corner leads into the major engineering feature of the canal, the deep Albert Street Cutting between high walls lined in the New Red Sandstone of the area. The overbearing nature of this cleft, with its solid bridges crossing, is enhanced by a number of heavy timbers between opposite walls, a quotation carved on each one.

Bridgwater Dock was home to *Teignmouth Electron*, found in the south Atlantic in 1969 with a falsified log but no sign of sailor Donald Crowhurst during the Golden Globe, the world's first round-the-world race.

The Blake Museum flies the Commonwealth Jack, 150 years older than the Union Flag. Admiral Sir Robert Blake was born here in 1599, becoming a general then an admiral for Cromwell. The museum features Blake and the Battle of Sedgemoor. The Admiral's Landing and residential and amenity building are all around and there is a large crane for lifting out cruisers. Old cranes and bases are still in evidence. A conical red-and-white object on the quayside is a Bristol Channel buoy from 1860, standing upside down for stability and exhibited as a feature.

Crowpill Coalyard was Brunel's last broad-gauge railway site, redeveloped for housing in 1995. A brick bottle kiln standing on the far side of the river is the last remaining of a forest of them that once made bricks and tiles from the silt in the River Parrett. These were subsequently sent as far afield as the West Indies. A double-leaf bascule bridge gives access to the smaller lower basin. This is overlooked by Russell Place, a row of Georgian cottages built in 1841 for dock workers, with a double-fronted villa at each end of the row for senior staff. In 1873, at the height of its prosperity, Bridgwater was Britain's fifth most important coal-importing port.

The canal's main engineering work, the imposing cutting at Bridgewater. All of the timbers bear industrial inscriptions.

20 Gloucester & Sharpness Canal

The Gloucester & Sharpness Canal or Gloucester & Berkeley Canal was built to bypass the uncertain conditions of the upper Severn estuary for shipping intending to travel upriver, although it stopped short of Berkeley, its proposed destination. Designed by Robert Mylne to take 1,000t ships, its construction began in 1794 but it was not completed until 1827, by which time the financial assistance of the Government had been enlisted. When finished, it was the widest and deepest canal in the world; the last to run a commercial passenger transport service, the Gloucester & Berkeley Steam Packet Company, which stopped operating in 1935.

The canal is all at one level with lock gates feeding from the River Severn into the docks at Gloucester, until recent decades the lowest crossing point of the Severn. The docks, Britain's best Victorian inland port, were completed before the canal so that access was only available at first from the river at the north end, locking up as the docks are normally 3.7m above river level. The current lock is 63m x 6.7m, able to take 400t vessels. The dock sides are high, often with lips that have been worn to a round profile. Dry docks are still in use. The custom house

Llanthony Warehouse and the Gloucester Waterways Museum.

has the Soldiers of Gloucestershire Museum. The docks are dominated by a number of seven-storey, 19th century, brick warehouses, which present a lofty atmosphere with an attraction that newer concrete structures lack. Llanthony Warehouse is home to Gloucester Waterways Museum with much canal memorabilia, amongst which a ghost was photographed in 2008.

Another warehouse contains the Gloucester Antiques Centre with over 100 dealers, one of the largest and longest-established in the UK. The Towpath Trail runs the length of the canal. Amongst moored craft is the former Spurn lightship *Sula*.

The Llanthony bascule bridge crosses over the canal. There are no fixed bridges on the canal; most are swing bridges to allow for high superstructures and masts. All traffic on the canal is controlled by traffic lights. Craft normally hoot to draw the attention of the bridge operator.

As the canal leads off down the Vale of Gloucester, following the River Severn in a south-westerly direction, it takes the user head-on into the prevailing wind. This is given free rein by the generally thin spread of trees and the flatness of the vale, which tends to funnel in towards Gloucester, intensifying the wind.

The canal breaks out into open country. The scarp slope of the Cotswolds moves steadily nearer on the left. The more distant Forest of Dean does the same on the other side of the river. Even the Malverns are visible at times, away to the north.

At Lower Rea is the first of the lock keepers' cottages. These are completely unique to this canal, built in classical Adam Regency style with pedimented porticoes on Doric columns, all differing slightly in detail and resembling so many Greek temples – an unexpected find on what was a very commercial waterway.

The landscape becomes totally rural, cutting through farmland that is only broken after the first of another grouping of high power lines.

Distance
27km from Gloucester to Sharpness
Highlights
Gloucester Docks and Waterways Museum
Lock keepers' cottages
Slimbridge Wetland Centre
Purton Hulks, Britain's largest ship graveyard
Sharpness Docks and Lock
Navigation Authority
Canal & River Trust
OS 1:50,000 Sheet
162 Gloucester & Forest of Dean

The canal crosses the River Frome with Wyecliffe College Boat Club alongside, followed immediately by Saul Junction, an unusual canal crossroads with the Stroudwater Canal. The older waterway was abandoned in 1954 and the remaining section is used for moorings. In its heyday it connected the River Severn via the Thames & Severn Canal with the River Thames at Lechlade. The Cotswold Canals Trust base is here and restoration is taking place eastwards, although not to reconnect the canal to the river, considered unsafe.

Frampton on Severn is a village constructed around Rosamund's Green, which at 800m x 100m is one of Britain's largest, including three ponds and a cricket pitch and surrounded by Georgian and half-timbered houses. It was named after Henry II's Fair Rosamunde, who was

The canal can handle large trip boats.

born at Frampton Manor. Frampton Court is a Palladian manor of 1733 to which a 1745 Gothic orangery and an octagonal dovecote were added. A Roman villa had a great mosaic and, at Church End, the 14th century canalside St Mary's church has excellent stained glass and monuments, delicate detailing to the turrets on the top of the tower and a Romanesque lead font.

The Tudor Arms stands by the canal at Shepherd's Patch, an area once known for its sheep grazing. Towpath walkers with dogs give way to walkers with field glasses, drawing attention to the headquarters of the Wildfowl & Wetlands Trust, although only the odd observation point of the Slimbridge centre can be seen from the canal. Established in 1946 by Sir Peter Scott, who also played a central role in setting up the Inland Waterways Association, Slimbridge has 160 resident species in addition to the migrant population – the world's largest collection – and is a 40ha centre for much important bird research. Varieties range from pink flamingos to trumpeter swans. These birds generally prefer not to use the canal. Canoe safaris allow a closer look.

Although the canal follows the River Severn closely, there is only one point where it is clearly visible from the canal, beyond where an overflow weir allows excess water to discharge into the estuary immediately alongside. Here, there are fine views across Waveridge Sand to the Forest of Dean, which is now close on the opposite shore.

Its embankment was shored up with 81 sunken wood, steel and concrete vessels, 30 of which are still visible. The largest ship graveyard in the British Isles, some of these Purton Hulks are of historical value and they have each been supplied by the Friends of Purton with a small plaque giving its details; an excellent, informal, open-air exhibition. One was an IRA gun runner until it was captured in 1921 by the Royal Navy. Seven of the 39 stemhead ferroconcrete barges are here, Britain's largest collection, precursors for the Mulberry Harbour units. *King* is a rare composite wood and steel vessel that shows how construction methods changed. Kennet barge *Harriett* is a scheduled ancient monument but most boats have no protection at all, unlike every blade of grass in the surrounding Site of Special Scientific Interest.

Ponds on the left of the canal were used at one time for storing complete tree trunks, keeping them afloat until they were required by the various timberyards.

The canal divides at Sharpness. The arm to the right, now used for moorings below a wall that is the remains of a coal tip loader adjacent to former pleasure grounds, was the original connection with the estuary, the end basin having been tidal. From 1939 to 1966 the TS *Vindicatrix* was moored here to train young men for the merchant navy, thousands of seamen identifying themselves as Vindi Boys. The lock is now disused beside the listed dock

Gloucester Docks with its warehouses, including the one that houses the Gloucester Waterways Museum.

keeper's cottage of 1856, used as a rescue base, and all traffic is routed through the docks.

These docks, covering 12ha of water, handle 500,000 tonnes of freight annually and are one of the leading sources of revenue for the Canal & River Trust's Freight Services Division. Timber has been a major commodity, the first timberyard belonging to the Canal & River Trust, but there are also more of the seven-storey brick warehouses with concrete silos reaching up to twice their height. A dry dock is available and narrowboats are constructed although these look totally lost among the shipping moored here. Shipping line agents occupy portable cabins and even the Customs & Excise office looks transitory.

The final lock down to the River Severn is one of the wonders of the canal network. The tidal range here is one of the largest in the world, the river drying out around the dock entrance at low tide so the lock is used only on the upper part of the tide. Not content with this, the lock adjoins a tidal basin some 200m long and 100m wide with mooring fingers projecting out from the sides. It is capable of holding ships up to 5,000t.

There is a magnificent view down the estuary to the Severn Bridges silhouetted on the horizon.

Gloucester

Hempsted

Quegeley

Gloucester & Sharpness Canal

Hardwicke

Saul

Saul Jn

M5

R Severn

Frampton on Severn

R Cam

Cambridge

Purton

Slimbridge

Sharpness

Berkeley

21 Monmouthshire & Brecon Canal

The Monmouthshire & Brecon Canal runs south-east across Powys from Brecon to Newport, a narrow contour canal followed by one of the best towpaths in the country. Two features that distinguish Welsh canals are their beauty and their proliferation of names, both being applicable in this case, not least because the canal started off as two separate canals. At the lower end was the Monmouthshire

Canal, built in 1796 to Pontnewynydd. The Brecon Canal, Brecknock & Abergavenny Canal or Brecon & Abergavenny Canal was to be built to the River Usk at Caerleon but the two were, instead, connected at Pontypool, a fact that was to save the upper canal in later years as a water feeder for the lower canal. Designed by Thomas Dadford Jr, it was opened throughout in 1812 to export iron ore, coal and limestone, but was also used for local coal, lime and cattle transport for farmers in the rural upper section. A plan for an extension to Hay was dropped in 1793 and the proposal for a canal from Abergavenny to Hereford was blocked by landowners. The upper canal was bought by the Monmouthshire Canal Company in 1865 to safeguard water supplies. There was no significant freight after 1870 and the canal was bought by the Great Western Railway in 1880. Tolled trade continued until 1938, although there were boat movements during the war.

The upper canal lies within the Brecon Beacons National Park, the requirements of which have resulted in it remaining beautiful. Restoration of this section was completed by British Waterways and Monmouthshire and Brecon County Councils in 1970 and the change of name recognises their efforts.

The gauge at the mouth of Ashford Tunnel gives an indication of how far it has dipped in the middle.

The canal is supplied by a feeder from the River Usk in Brecon, a town built at the confluence with the Afon Honddu and Afon Tarell. Mostly 18th century, the town has Roman origins. Brecon Castle is largely 11th century, with the remains of a large motte, bailey, walls and two towers. Most of the destruction took place in the Civil War with the assistance of the locals, who were shrewd enough to realise that they would have a quieter life if neither side occupied it. The Priory Church of St John, founded as a Benedictine priory in 1093, is mostly 13th century but with a 14th century nave. It is noted for its fine glass and side chapels dedicated to medieval trade guilds and has the colours rescued from Rorke's Drift in 1879. It was made a cathedral in 1923. The Brecknock Museum includes the dugout canoe from Llangorse Lake crannog, the most famous Welsh dugout, some 1,100–1,200

Distance
66km from Brecon to Newport
Highlights
Brecon Beacons National Park
Seventeenth century Grade II Llangynidr bridge
Navigation Authorities
Torfaen Council/ Newport City Council
Canal Society
Monmouthshire Brecon & Abergavenny Canal Trust
www.mbact.org.uk
OS 1:50,000 Sheets
160 Brecon Beacons
161 Black Mountains
171 Cardiff & Newport

Residents on the Tycock lock flight.

years old, described in *Dead Men's Boats*. It may have belonged to Brychan, the Welsh king who gave his name to the town and whose Irish father, Anlach, would have been associated with canoes and crannogs. The Regimental Museum of the Royal Welsh features a Zulu war room.

The canal slips away quietly past the backs of small gardens, forming the southern edge of Brecon. The road that the left bank then follows was a Roman road and later the A40. Traffic now bypasses the city on the far bank of the River Usk, which will never be too far away until Abergavenny. Hawthorns and occasional low-hanging willow and ash trees give the canal very much the feel of a river; the *Dragonfly* trip boat passes regularly so navigation must be easier than it appears. The trees obscure views of Slwch Tump fort, a hilltop aerial and barracks on the north side. The view is dominated by a wooded ridge to the south. Over the length of the canal the prevailing wind is either following or is blocked by the mountains, which provide imposing scenery to the south. The valley ahead gives extensive views to the east.

The canal runs along the right side of the valley towards Pontymoel with the towpath on the left bank. Bridges have been built with the arch higher over the towpath than over the canal. In addition, the right side of any road across is frequently much higher than the left so that the asymmetric arch has a point of weakness above the towpath and many of the bridges have had their arches reinforced to counter this problem.

Near the Star Inn the canal reaches a low electrically operated drawbridge of 1970, with a notice forbidding it from being opened during rush hours for the local school. The canal passes houses at rooftop level to cross the Caerfanell aqueduct, the embankment having been conspicuously lined with concrete placed around the trunks of all the trees.

The 343m Ashford Tunnel has no towpath so boats were pushed through with poles using notches in the walls. The tunnel dips in the middle so there are height marker boards across the tops of the tunnel mouths. A vertical shaft in the centre adds a welcome but initially puzzling

pool of light. A chain hangs along the side to give assistance if required.

Cwm Crawnon Lock and the four Llangynidr Locks drop the canal to a 40km level pound, the longest in Wales, despite the mountainous terrain. Among the beechwoods, wild cherries, bracken and foxgloves are a stone warehouse used by Country Craft Narrowboats, a Canal & River Trust depot, a picnic area and a wooden roller next to the Afon Crawnon Aqueduct for removing the drain plug.

The Coach & Horses Tavern is noted for its fine French cuisine and dining tables are often placed placed on the bank above the canal, 100m before the last lock.

The canal is crossed by the B4560 at Coed-Yr-Ynys, the road then crossing the river at Llangynidr bridge, a 17th century Grade II structure with six arches divided by triangular cutwaters and pedestrian refuges.

Although the higher mountains have been to the south, a few of the hills on the opposite side of the valley have been prominent, including Allt

The boat hire centre at Llanfoist.

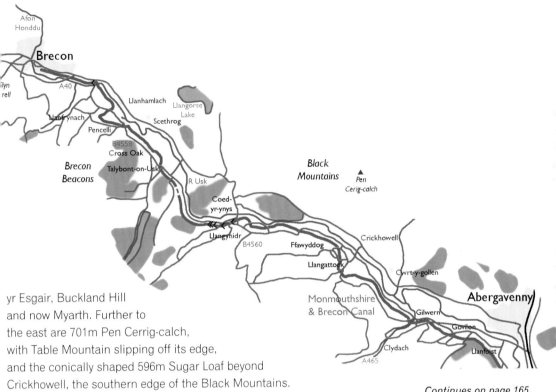

yr Esgair, Buckland Hill
and now Myarth. Further to
the east are 701m Pen Cerrig-calch,
with Table Mountain slipping off its edge,
and the conically shaped 596m Sugar Loaf beyond
Crickhowell, the southern edge of the Black Mountains.
Myarth has a standing stone at each end and conspicuously out-of-place
Italianate towers in orange brick at Gliffaes.

Continues on page 165.

Crickhowell, with its 18th and 19th century houses, is reached via a
medieval 13-arched stone bridge over the river, guarded by the Norman Ailsby's
Castle. This was destroyed in the 15th century except for the motte, bailey,
curtain wall and small tower, which partly survive. The 14th century church has
stained glass and other local features include the Porth Mawr gatehouse and a
burial chamber. The town was formerly a centre for Welsh flannel production.

There is another standing stone across the river by Cwrt y Gollen army
camp, from where the noise of rapid gunfire might be heard.

The Bridgend Inn and Navigation Inn stand at opposite sides of the canal
beyond the Gilwern Aqueduct, which clears the River Clydach in a single
arch. Gilwern had two tramroads, one passing under the canal and along
the Clydach valley, to bring down coal from the Valleys and pig-iron from the
Clydach gorge ironworks to Gilwern Wharf, now a picnic site. A sign points up
the hillside to the Lion Hotel, in the direction of the A465 Heads of the Valleys
Road, which now comes alongside for a noisy kilometre until it crosses and
moves away. An aerial on Gilwern Hill acts as a landmark but from close up it
is hidden by the trees.

An oblique, disused railway bridge at Govilon has been an opportunity
for the bricklayers to demonstrate their skill with heavily skewed courses.

The portal of Cwmbran Tunnel hides in the shadows.

It now carries a Sustrans cycleway. Govilon Wharf formerly had an ironworks, limekilns and tramway. Now it is the home of Govilon Boat Club, the first inland cruising club in the country, dating from 1963.

From Govilon, the Monmouthshire & Brecon Canal runs south. The canal forms the eastern boundary of the Brecon Beacons National Park as far as Pontymoel.

Beyond the trees lies the gateway to Wales, Abergavenny (the Gafenni being the blacksmith's river), with the Ysgyryd Fawr and the Roman site of Gobannium on the Ysgyryd Fach. The 11th century Abergavenny Castle is built on a mound. In 1177, William de Braose invited the notable Welsh leaders to dine and then, suddenly, had them all murdered in order to secure control of the surrounding area.

Llanfoist Wharf is home to Beacon Park Boats' hire centre, which is arguably the most attractive in the country; it has no advertising hoardings, just lawns, rambling roses and stacks of logs for

the winter in a beautiful setting. Views over the border country are extensive. It is hard to believe that this rural spot is only 4km from the Welsh pit village of Blaenavon.

The mountains begin to pull back at Goetre. There is a spur to the right that served Goytre Wharf with its limekilns, now home of Red Line Boats, a boat hire marina. The Waterside Rest offers meals and displays with information on the local industrial heritage.

Pontypool, Welsh for 'the bridge at the pool', stands on the rim of the coalfield and has been industrial since Roman times. In 1720 it produced the first tinplate in Britain and in the 18th and 19th centuries it was an iron town. It was a japanning centre in the 19th century but also a market town. The Italianate Grade II town hall of 1856 was donated by ironmaster Capel Hanbury Leigh, to celebrate the birth of his only son. Pontypool Park House was built for the Hanbury ironmaster family and is now a school – St Alban's RC High School – for 11-to-18-year-olds. Its Georgian stable block houses the

Valleys Inheritance Centre. The wrought-iron entrance gates were given to John Hanbury by Sarah Churchill, Duchess of Marlborough.

For many years the navigation ended at a 2.4m diameter culvert at Sebastopol. Crown Bridge has been rebuilt and powered boats are now able to continue along a broad reach to Five Locks.

The 80m Cwmbran Tunnel, with no towpath, lies below a golf course. The water after the far portal is shallow at first but deepens towards Pontnewydd Locks. Five Locks Moorings were opened in 1998. Restoration of the rest of the canal is under way. The canal fell 103m in 14km through 31 locks.

Below Five Locks, the canal is controlled by anglers who claim that any boats other than their own on the navigation create disturbance and so are not permitted.

As a landscaped water garden, the ten locks of the Cwmbran Flight begin very attractively. As a canal they have less to recommend them. The long walk down to Forge Hammer passes the Old and New Bridgend Inns and a selection of modern industrial units. Launching is possible among the shopping trolleys but the water comes to an end again where 500m of canal has been covered by a roundabout and a busy road. A vertical cutting for the road means that the line of the canal cannot even be followed on foot. Instead, a diversion is required up the footpath to the right from the roundabout, along a residential road, left past some shops and right at the bridge crossing the new road.

The canal begins again where an iron beam from the Halfway Bridge of 1847 has been erected across the end of the reach as a feature. A swan's nest has been visible at a surprising site in the middle of a housing estate with only 300m of clear water. A low bridge crosses halfway down this reach. Another short walk is required where a housing estate road has been laid along the line of the canal. Further walking is needed at the Waterloo Lock, two Oakfield Locks and then, from Ty-côch, ten locks dropping away in open country.

The canal comes to a sudden end at the ivy-covered Barrack Hill Tunnel, where a screen of mesh confronts the boater and the noise of water can be heard tumbling away into the darkness. The canal used to serve Newport Docks but sections were closed in 1879 and 1930, making any further restoration extremely difficult.

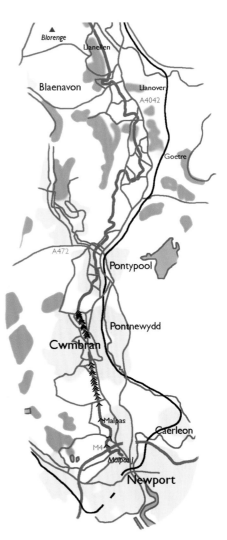

Stratford-upon-Avon Canal

22

The Stratford-upon-Avon Canal has often been considered as two different canals. The northern section was opened in 1802 to provide an alternative route from the Grand Union Canal into southern Birmingham by linking across to the Worcester & Birmingham Canal at King's Norton. The resulting canal also happens to be more attractive than the Grand Union Canal.

At King's Norton a turn off the Worcester & Birmingham brings a pair of fine guillotine stop gates, now rusted solid but originally designed to be dropped to save water in the event of a breach in either canal. The line of a former swing

Edstone Aqueduct is the longest in England.

bridge was the scene of the first IWA protest cruise by Sir Peter Scott and LTC Rolt when the bridge was fixed by the GWR but they were obliged to jack it up to allow navigation passage.

The only tunnel is at Brandwood. Its 320m length is barred to unpowered craft.

In Birmingham, the Horseshoe public house is adjacent to an arm that is approached under a brick arch bridge. Many of the older bridges on the canal are of red brick with surprisingly large arches. But one or two of the newer concrete structures have lengths of rail at the corners to prevent towropes rubbing and have raised lines of brickwork on the towpath that once allowed horses to grip and now thwart cyclists.

In 1936 the Government tested a bridge at Yardley to destruction. The 150-year-old structure held 125t for 45 minutes before collapsing.

The Blue Bell Cider House at Waring's Green has had a long history as a public house and also as farm buildings, the publican supplementing his income by selling farm produce to the canal folk.

At Lapworth is a bascule bridge that gives access to a farm, reminiscent of those on the

Working the Kingswood Flight.

Distance
41km from King's Norton Junction to the River Avon
Highlights
Bancroft Basin
Royal Shakespeare Theatre
Navigation Authority
Canal & River Trust
Canal Society
Stratford upon Avon Canal Society
www.stratfordcanalsociety.org.uk
OS 1:50,000 Sheets
139 Birmingham & Wolverhampton
151 Stratford-upon-Avon

Oxford Canal. The end of the top pound is reached with the first of the 55 locks down to Stratford. Although some are in flights, especially at Lapworth, the spacing, typically, is at about 400m intervals.

A kilometre to the left of one of the bridges stands the timber-framed Packwood House. Built about 1560, it is opened by the National Trust and contains interesting tapestry, needlework and furniture. More notable, perhaps, is its fine

Carolean garden and yew garden clipped to represent the Sermon on the Mount.

Two bridges over the Lapworth Flight at Kingswood are the first of many split bridges on the canal. These have slots in the centre to allow the towrope through without unhitching the horse and are built as pairs of cantilevered arms, the handrailing being in the form of a distinctive double cross on each side. The locks themselves have large side pounds between them, so large that some are used as moorings. They have small paddles, making them slow to operate. Just below the Boot public house and a nearby grocer's shop, one has an overflow that discharges a cascade 2m high and nearly as wide, rather than having the usual discharge channel running through the undergrowth around the back of the lock. There is also some ornate footbridge ironwork to be seen.

Despite the fact that the two run parallel for some distance, the Stratford-upon-Avon falls rapidly through a whole flight of locks while the Grand Union is level from Rotton Row to Hatton, a distance of some 14km.

The southern section, linking up with the River Avon at Stratford, was completed in 1816, 14 years after the northern section. Unusually,

The Bard in front of Stratford basin.

it has been owned by the National Trust and its recovery from disuse was a personal triumph for David Hutchings. Volunteers, services and prison labour restored it to use in three years and it was reopened by the Queen Mother in 1964, a model for other canal restoration schemes and the first restoration led by volunteers.

Immediately, an attractive set of canal buildings face the first lock of the southern section while just below, on the other bank, stands the first of half a dozen houses unique to this canal. They have barrel roofs, each house being built like a single bridge arch with ends. A couple of them now act as canal gift shops.

Above the house is a cut to the short Lapworth Link to the Grand Union Canal, added to prevent boats coming up from Stratford and heading for the Grand Union needing to lock up to Kingswood Junction and immediately down again to the connecting arm, which would be a waste of time and two locks of water.

Yarningale Common has a short cast-iron aqueduct over a stream. Like the two other aqueducts to follow but unlike almost any others on the canal network, the towpath is at the level of the invert of the trough so that the horse had to climb down on to the aqueduct towpath and up off it at the other end. Because it leads to Bucket Lock, canoeists must portage on to the towpath, one of the few occasions they will ever have to climb down out of their boats.

The Anglo Welsh canal basin and boatyard at Wootton Wawen, formerly a temporary terminus, is served by the Navigation Inn. Just visible over the trees down the road is the roof of the magnificent five-storey 18th century brick watermill on the River Alne. Beyond it stands Wootton Hall and then St Peter's church with a Saxon crypt. It is the oldest in the county and one of the oldest in England. It also boasts an 11th century sanctuary, a Norman nave, a superb 14th century east window and a lady chapel like a barn.

Up the hill, Austy Manor can be seen and, further beyond it, Austy Wood, the largest remaining section of the Forest of Arden that at one point covered the whole area.

Wilmcote is visited by tourists seeking what was incorrectly believed to have been the Tudor house of Mary Arden, Shakespeare's mother, and its attendant farming museum and cider mill.

Stratford is approached the back way, past industrial estates. Narrowboats are moored at Valley Cruises. A cafe stands on the A3400.

To most people Stratford is Shakespeare. His birthplace lies just beyond. Stratford has more historic buildings than any other town of comparable size in the country and simply must be visited.

Bollards carry the names of benefactors of the canal in the area that is now residential, some of it recent, as the canal wanders near the back of the bus station and the Red Lion before bursting out into the splendour of the Bancroft Gardens, set around the canal basin. On one side, the Bard sits on a pedestal, surrounded by flowerbeds and statues of his characters. Opposite is the Royal Shakespeare Theatre, built in 1926 in dubious cinema architecture to replace the 1879 one destroyed by fire. The Royal Shakespeare Company are in residence from April to December. The tourists are always in residence, especially in the summer, and all the facilities here face the gardens, from McDonalds to a venue that offers visitors the chance to sample life in Shakespeare's day.

Significant tourist dates are April 23rd, Shakespeare's birthday, and October 12th, Mop Fair, the time when farm workers were traditionally hired and which is still celebrated.

The final lock down to the River Avon is a broad-beam one. Originally, the whole canal was to have been broad gauge but when the Worcester & Birmingham Canal and Warwickshire canals were constructed as narrow canals this idea was abandoned.

Downstream the 15th century Holy Trinity church in which Shakespeare is buried can be seen.

23 Worcester & Birmingham Canal

Gas Street Basin in the centre of Birmingham might be considered the central point of the British canal network. This was certainly the case in 1791 when the Worcester & Birmingham Canal Act was passed. The Staffordshire & Worcestershire Canal Company obtained an injunction to prevent the Worcester & Birmingham coming within 2.1m of the Birmingham Canal Navigations, resulting in the Worcester Bar, over which goods had to be manhandled in an effort to prevent loss of business and water to the new canal. When the Worcester & Birmingham was opened to traffic in 1815, a stop lock was opened through the bar and this provides a route through with a footbridge over the top.

The Worcester & Birmingham was acquired by the Sharpness New Docks Company in 1874, nationalised in 1948 and kept in use by Cadburys until 1964.

The top end of the Tardebigge Flight of locks.

Edgbaston Tunnel is just 96m long and has a towpath. The railway goes through an adjacent bore and a heavy arched retaining wall flanks it at each end. The other bank comprises red sandstone and supports Birmingham University campus, with its brick clocktower landmark in the centre and silver domed roofs.

At Bournville, in 1879, the Cadbury brothers set up their chocolate factory as a garden city on what was then a rural canalside site. This pioneer community now has 3,500 houses. Old wharves front the canal and a plank footbridge, lifted by a crank handle, stands sentinel on the left bank. A housing estate occupies Cadbury's former wharf, opposite the factory, sold to Kraft Foods in 2010.

Wast (or West) Hills tunnel is the major obstacle on the canal. At 2.3km long it is one of the longest still in use in Britain. The tunnel emerges in deep cutting in open countryside, the city having been left behind.

At Arrowfield Top is the Hopwood House Inn. Barnt Green sees boats moored alongside Lower Bittell Reservoir, built to compensate mill owners for their loss of water after the construction of Upper Bittell Reservoir, which is a canal feeder. The reservoirs are popular with anglers and birdwatchers but there are few anglers on the canal itself.

Houses here have large gardens beside the canal. Wheeley Farm overlooks the canal on top of its hill, which is steeper than most in this part of the country. The Crown is passed and the small Weighbridge offers real ale and food behind Alvechurch marina.

Shortwood Tunnel is 560m long. It is possible to see daylight dimly through this tunnel and the following one, the 520m Tardebigge tunnel.

Natural ventilation by the wind causes exhaust to rise silently from the downwind ends of the tunnel, an eerie sight in the evening light. Shortwood Tunnel was reputed to be haunted and to have an archway below water level,

leading to a secret cavern, although this has not been found. Tardebigge tunnel has moorings at both ends and points straight towards the spire of the church in Tardebigge.

The land starts to drop away as the top lock of the Grade II Tardebigge Flight is reached, marking the end of the 27km level pound from Birmingham. It was at this lock that Robert and Ray Aickman met Tom and Angela Rolt on *Cressy* in 1945, which eventually led to the setting up of the Inland Waterways Association.

The canal and the railway pass the university campus.

Distance
48km from the Birmingham Canal Navigations to the River Severn
Highlights
Tardebigge lock flight
The Mailbox and the Cube
Navigation Authority
Canal & River Trust
Canal Society
Worcester, Birmingham & Droitwich Canals Society
www.wbdcs.org.uk
OS 1:50,000 Sheets
139 Birmingham & Wolverhampton
150 Worcester & the Malverns

The Commandery played a central role in the Civil War.

This flight is the longest in the country, dropping through 30 locks in the next 3.4km, with a further 28 to go down to the River Severn. Some of the locks have hooks at their lower ends to assist horse towlines.

The top lock is one of the deepest narrow locks in the country with a fall of some 4m. It was originally built as a lift lock but was converted to a conventional lock in 1815 after it was found not to work properly in its original guise.

It takes a narrowboat about four hours to lock through the first flight but there can be few more pleasant places to pass the time if the weather is right. Apart from the odd lock cottage, an array of radio ham aerials and a distant farm, there is nothing to break the solitude and beauty of the countryside. The canal drops past hawthorn bushes and fields of hay and barley.

The Queen's Head Inn, on the right, is a public house with a substantial carvery.

The lock cottage at Stoke Pound is hired out by the Landmark Trust, one visitor suggesting that sitting in the cottage and watching the boats go past was a better activity than jumping on and off a boat and watching the cottages go past. Hardly has the Tardebigge Flight been left when the six locks of the Stoke Flight are encountered, leading down to the Navigation Inn, a large polyethylene-covered boathouse and Black Prince's base.

The Boat & Railway, now offering a carvery instead of its former famous pickled eggs, draws attention to the high-speed trains running just west of the canal.

Surprisingly few orchards can be seen, considering the part of the country concerned, but one is passed on the left and it is followed on the right by low ground giving views right across to the Malvern Hills. The cream block of Hadzor Hall dominates the hill on the right.

The Droitwich Junction Canal branches off and the Mid-Worcestershire Ring is met. The Eagle & Sun offers food on the line of the Roman Salt Way. On the other bank, a lighthouse marks Hanbury Marina.

At Oddingley, a squat stone church overlooks a fine half-timbered building, forming a very attractive group. The vicar was murdered in the 19th century for raising the tithes, the murderer being killed in turn to preserve silence.

The canal has been of generous width for much of its length but, after Tibberton and the Bridge Inn, it achieves river-like proportions and passes through a deep cutting as it approaches the M5, an old barn and the six locks of the Offerton Flight, overlooked by the cream bulk of the house at Hindlip Park.

Increasing use is made of galvanised sheet piling, which gives the canal a well-kept look. Tollandine Lock allows the first view of the northern outskirts of Worcester, a city that has been noted for gloves since the 13th century, Royal Worcester Porcelain since 1751 and Worcestershire sauce since the 1820s. After Blackpole Lock comes a substation and an industrial area, followed by extensive playing fields and a sports centre on the right side.

The Cavalier Tavern offers food and coffee with views to a squat, towerless brick church one way and the architecturally dramatic brick Rainbow Hill railway bridge over the canal the other.

Blockhouse Lock gives a close view of Worcester Cathedral, begun in 680, the current building started in 1084 but damaged in the Civil War. Its tomb of King John is said to be England's oldest royal effigy. Kings Head Lock is alongside fine half-timbered tea rooms. The 15th century Grade I Commandery – built on a Saxon chapel site that was used for a hospital in the 11th century by St Wulfstan until dissolved in 1540 – was also used by Charles II as a council room before the Battle of Worcester, has been a glove workshop and now has a visitor centre. The bridge across past the Kings Head is edged with sculptures of Civil War helmets and lances. The Tudor House Museum with 1910 bathroom and Edwardian schoolroom and the City Museum & Art Gallery are near the 19th century Shire Hall, while the brick Guildhall dates from 1721.

The canal skirts almost unnoticed round most of Worcester and slinks into Diglis Basin the back way, offering the Anchor Inn and the Cafe Afloat in an area through to the river that has been extensively redeveloped. Housing has replaced industrial premises. The final two locks down to the River Severn are both broad locks.

24 Droitwich Barge Canal

Droitwich has long been important for its salt. The Romans used the River Salwarpe to export it. The river was poorly canalized from 1662. The parallel Droitwich Barge Canal by Brindley was opened in 1771 as a wide beam canal to take trows working off the River Severn.

The narrow Droitwich Junction Canal by Boddington, connecting with the Worcester & Birmingham Canal, was not opened until 1854, after the railways

The new staircase lock.

The Droitwich Junction Canal leaves right at Hanbury Junction.

Distance
12km from the Worcester & Birmingham Canal to the River Severn
Highlights
Vines Park and brine well
Chateau Impney hotel
Navigation Authority
Canal & River Trust
Canal Society
Worcester, Birmingham & Droitwich Canals Society
www.wbdcs.org.uk
OS 1:50,000 Sheets
(138 Kidderminster & Wyre Forest)
150 Worcester & the Malverns

had started. It was all abandoned in 1939. The Droitwich Canals Trust was formed in 1973 and was faced with some interesting technical problems. Full reopening of the canal took place in 2011, offering the Mid Worcestershire Ring as a convenient and pleasant 34km weekend circuit for cruisers.

The Droitwich Junction Canal leaves the Worcester & Birmingham Canal at Hanbury Junction by Westfield Bridge and the Eagle & Sun and follows the B4090, the Salt Way Roman road from Alcester.

Use by the army as an equipment test area resulted in the flight of three Hanbury locks being in good condition, some of the last narrow locks built in Britain. They each fall about 3.7m and have brick side ponds, which have been restored and are in use.

The cream Hadzor Hall and adjacent church look down from a hilltop on to a 1km diversion resulting from housing being built on the former line of the canal. A corrugated culvert carries a road that passes the Gateway Park and serves a new 240-berth marina and Droitwich Rugby Club. Two new wildlife ponds have been dug and the wildlife includes mallards, moorhens, frogs, great crested newts, grass snakes and slow worms. Brackish pools encourage the growth of wild celery and lesser sea spurrey.

On a ridge to the north is the Chateau Impney, built in 1875 in Louis XIII style by salt mine and canal fleet owner John Corbett for his wife, Anna Eliza. It is now a hotel and conference centre.

The Barge Lock takes craft off the river above a small but aggressive weir on to the Droitwich Barge Canal, built broad to take sailing vessels, and which has a swing bridge across the chamber. The barges were highlighted in red and green, making them colourful, and carried sails.

St Augustine's church on Dodderhill is the most prominent building locally, partly Norman.

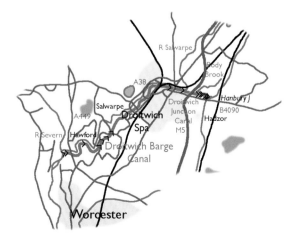

Below in the park is a statue of St Richard of Droitwich, who declined inheritance of an estate and went on to become Bishop of Chichester with a simple lifestyle.

The town was called Wich until Edward III formally gave the right, le droit, to manufacture salt and this was added to the name. The Spa postscript was also added as this is Britain's only saltwater spa town, developed by the Victorians from 1836 as a fashionable watering place. Droitwich Spa Brine Baths were revived in the 1980s with water ten times as salty as seawater or 40 per cent more salty than the Dead Sea.

The 19th century Chateau Impney, now a hotel and conference centre.

The source is rock salt deposits 60m below the town. This is considered to be Britain's highest-purity salt and has produced a third of Britain's needs, especially during the Hundred Years War when imports were cut off. A brine well, enclosed by ornate railings, is next to the canal in the park, opposite a Waitrose store.

The Gardners Arms is close to the park but the best of this market town with its timber-framed houses stands away from the canal. Droitwich Spa Heritage & Information Centre has a museum of local history featuring the Romans and salt extraction plus a brass-rubbing centre. Droitwich Music & Arts Festival has conflicted with canal festival plans.

The Netherwich Basin was dug to load salt on to some 50 Wych barges. Marina users are faced by a slipway. Damage to canal structures has resulted from subsidence and the salty canal water is not good for steel hulls but has helped to preserve wooden structures.

The Railway pub has railway memorabilia and a view across to the much-sprayed Droitwich Amateur Boxing Club premises. They stand by Netherwich Tunnel, carrying the New Street to Hereford railway and drawing in a branch from Kidderminster. The tunnel, more a large bridge than the typical canal tunnel, has bats and these are included in some silhouette art panels at the end.

The canal turns southwest past a sports centre, the Great Pool and a nature reserve. The A38 crosses over as it bypasses the town at Chawson where 120,000 reeds have been planted over 2.5ha of Coney Meadow. These are to compensate for restoration that has removed reeds that have grown across the canal. In fact, reeds have also been left to occupy 3m of canal edging so there is no shortage of this habitat, although a large quantity of chopped reeds float in the canal.

The rest of the route is rural. At Salwarpe the half-timbered 15th century Salwarpe Court and Old Mill House are passed. Salwarpe Bridge is one of the places on the canal that used early ball bearings. Brindley placed two rising floodgates in the bed of the canal, intended to slam shut in spate conditions. Siltation buried them, although they were preserved by the salty water.

The Monarch Way follows the canal, based on the escape route of Charles II after the Battle of Worcester in 1651.

The first Ladywood Lock has cranked balance beams as it is so close to Ladywood Bridge. The four Ladywood locks were tackled by 580 volunteers during the 1973 Droitwich Dig, pushing the restoration in the right direction

and sending reeds to thatch huts in the West Midlands Safari Park.

More powerlines cross between Porter's Mill Lock and Mildenham Mill Lock, where Brindley included a circular weir. To the north-west of the river is Hawford Dovecote, which still has many nesting boxes. The Grade II Linacre Bridge of local bricks and stone is built on elm tree trunks laid horizontally, probably Brindley's last remaining of the type left unmodified.

The wooded ridge at Hawford stands out prominently after what has been mostly gentle countryside. A campsite on the north of the canal precedes the A449 crossing, which had been lost. Plans to take the canal through the large stone arch over the river were thwarted, resulting in a costly construction exercise to install what is more like a tunnel than a bridge under this major dual carriageway.

The two Hawford locks take the canal down to the River Severn. The bottom one was Brindley's favourite construction, in the early days of canal building. Restoration has included installing tall piles with floating pontoons to handle the large fluctuations of level on the Severn. Such luxuries were not available to those trying to sail Wych barges into this canal.

Cranked balance beams at the first Ladywood Lock.

25 Staffordshire & Worcestershire Canal

Running southwards through the West Midlands, the Staffordshire & Worcestershire Canal was an early James Brindley canal, built 1768–1772 to join the Trent, Mersey and Severn as part of the Grand Cross scheme that was also to have a link with the Thames. It was one of the first canals of the Industrial Revolution to be completed and was the main route north from Birmingham until the Shropshire Union Canal opened. The Worcester & Birmingham also took away some traffic later on a more direct line to the Severn but it remained a considerable commercial success and was one of our most prosperous canals. It was constructed as a narrow-beam contour canal so views are constantly changing and it is frequently above the level of the surrounding fields.

The Staffordshire & Worcestershire Canal leaves the Trent & Mersey Canal at Great Haywood Junction, sandwiched between the River Trent and the Rugeley to Stoke-on-Trent railway. Anglo Welsh Waterway Holidays are situated in one of several narrowboat marinas that appear down this canal.

The towpath bridge over the junction has an unusually wide bridge arch of 11m with two courses of bricks overlain by sandstone slabs. All bridges on the canal are clearly named and numbered and the

The round house at Gailey Lock.

Roving bridge and stoplogs at Milford.

aqueducts say which rivers are being crossed. Trent Aqueduct in stone carries the canal over its river and another arch clears a mill tailrace as the canal starts up the valley of the River Sow. Great Haywood Wharf has mellow red-brick warehouses but the attention is drawn more to the tiny brick toll office on the other side of the canal with its iron-lattice windows, now used as a craft shop.

On the other side of the canal the large building with the long arched front is merely Tixall Farm but the restored four-storey arched gatehouse and stables of 1580 point to something grander. The first house was also built in 1580 by Sir Walter Aston and Mary, Queen of Scots was kept a prisoner here for a fortnight in 1586. In 1678 another Aston was implicated in the Titus Oates conspiracy.

Tixall Lock is the first of the dozen that lift the canal 30m to its top level at Gailey. From the lock can be seen the largest engineering structure on the Trent Valley railway line, the 708m Shugborough Tunnel curving under Shugborough Park to follow the line of the canal. In keeping with the locality, it has battlemented tower portals with additional towers at this end.

There is a parting of the ways as the canal turns southwards to follow the valley of the River Penk. The railway crosses to head for the centre of Stafford, as did the former Stafford branch of the canal, which continued up the River Sow until it was abandoned in the 1920s. Stafford's name comes from the Old English *staeth ford*, ford landing, when the waterway was shallower.

Junction 13 of the M6 is just before Acton Trussell. But for the traffic noise, the village retains its old charm with beehives beside the canal and the 15th century church looking down

Distance
37km from Great Haywood Junction to the River Severn
Highlights
Stourport-on-Severn, Britain's best canal town
Navigation Authority
Canal & River Trust
Canal Society
Staffordshire & Worcestershire Canal Society
www.swcanalsociety.co.uk
OS 1:50,000 Sheets
127 Stafford & Telford
138 Kidderminster & Wyre Forest
139 Birmingham & Wolverhampton

An impressive sandstone cutting near Wolverley.

across mown lawns to the Moat House Hotel with its bars and restaurants, which is partly 14th century timbered.

Rodbaston, Boggs, Brick Kiln and Gailey locks take the canal away from the motorway, and up to its 17km summit pound. The canal thus avoids Junction 12 of the motorway, which is surrounded by the Gailey Pools reservoirs that not only have a heronry but large flocks of ruddy ducks introduced from North America and now seen as pests by ornithologists in this country.

The road from the motorway to Gailey wharf is the A5, built by the Romans as Watling Street, here just short of Water Eaton, the Roman burga of Pennocrucium, together with a legionary fortress, two forts and two temporary camps.

At Slade Heath the railway line finally crosses and makes its departure. The A449, which has followed a similar line and is a Roman road here, passes close to the Fox & Anchor public house at Cross Green and then also crosses and departs in the direction of the railway. Not far away is Featherstone Prison.

The canal now passes under Junction 2 of the M54 and into a progressively more built-up section of Wolverhampton at Fordhouses. Although the worst excesses of the built environment are avoided there is little to show of the *hean tune*, the Old English for high village, of which the lady of the manor in 985 was Wulfrun.

Aldersley Junction quickly follows, the bottom end of the New Main Line of the Birmingham Canal, which climbs steeply up the side of Wolverhampton horse race course from where the pounding of hooves on turf may be heard.

This is now part of the Stourport Ring. A Norman church tower with battlements is hidden by the trees at Tettenhall where the A41 crosses. The oblique railway bridge beyond it no longer carries trains but the line between Oxley and Castlecroft has become the Valley Park nature trail and cycle route.

Compton Lock brings the end of the summit pound. This is where Brindley began building his canal and, until it was recently rebuilt, the lock was reputed to have been the oldest lock in the West Midlands.

Compton Wharf Lock is the first of 31 dropping the canal 81m down to the River Severn. Wightwick Lock looks across to the 29ha of Victorian/Edwardian gardens with topiary, yew hedges, terraces, ponds and woods that surround Wightwick Manor. Ornately timbered in Jacobean style, the house dates from 1887–1893, was furnished in Arts & Crafts style and contains Kempe stained glass, original William Morris wallpaper, tapestries and fabrics, paintings, de Morgan ware and other pre-Raphaelite works of art, and there are also stables, along with a pottery, studio workshop and antiquarian bookshop.

Many of the lock weirs on this canal are of an unusual design: a circular weir protected by a circular brick wall and a birdcage arrangement in the centre to keep out both people and debris. The example at Awbridge Lock is typical of a number of these weirs.

The three Bratch Locks were originally built as a staircase but extra gates were later added to give 1.5m gaps between the locks. A side pond curving away to the right between the closely spaced locks tempts the unwary although, thinking about it, there is no way narrowboats can reach it. The correct route is down the flight and through the delightfully complex brick

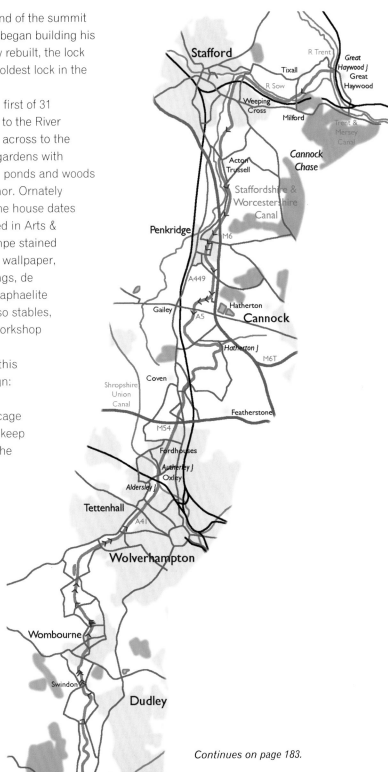

Continues on page 183.

access stairways at the bottom lock. Watching over everything is a strategically sited 18th century octagonal toll house. The area is one of the gems of the canal system, no less so for the restoration work carried out by British Waterways in 1994.

The Round Oak and the Waggon & Horses mark the two ends of the jig through Giggetty, the westward extension of Wombourne. There are also some large industrial units by the canal, giving way to a wildlife centre at Smestow before the two Botterham Locks.

What appears to be an isolated round weir opposite a section of wooded hillside well away from Gothersley Lock is the remains of Gothersley Roundhouse, rebuilt as a picnic area. Originally three storeys high with a rectangular extension, this tower was built in 1805 to house a wharfinger.

A cave cut into the red cliff, the Devil's Den, can really only be approached from the canal but at a point where the water is too shallow for large craft to get close.

The Upper Basin at Stourport and the decorative clocktower.

Below Dunsley Hall is Dunsley Tunnel, Britain's shortest official canal tunnel at 23m. Before Kinver Lock there is a large pumping station. The dominant building for the village, however, is the church high up on a ridge above the canal. The ridge extends westwards and runs south-west as the gorse- and heather-covered Kinver Edge. On the west side of the village is Holy Austin Rock where a series of caves had brick fronts built and were occupied from the 16th century by up to 12 families, some 80 people, until they were moved out by health officials in the 1960s. One is again occupied.

The lock cottage at Whittington Lock is also attractive but the most interesting building is the timbered Whittington Inn, the former manor house of 1300, up the hill on the left. At one stage it was owned by the grandfather of Dick Whittington, the boy who went to London and became Lord Mayor after he heard the streets were paved with gold. The resident ghost is that of Lady Jane Grey, who lived here. There is a 300m hidden tunnel to Whittington Hall and priestholes are built into it. A final odd building is the wooden shed on stilts on the embankment

after the lock, originally fitted with wheels that have not been removed but remain attached, along with the supporting piles.

Cookley or Debdale Lock has a large cave cut into the sandstone next to the lock chamber. Probably used as a store or keeper's shelter in the past, it has only a small doorway but could easily hold a couple of cars inside. The lock is followed by rock cutting with walls up to 6m high.

From Wolverley Court Lock with its formerly split cantilevered footbridge, the canal works its way towards Kidderminster – Old English for Cydela's monastery.

One of the better spots is just before Kidderminster Lock where a Perpendicular church tower stands on top of a bank of cultivated shrubs above a wharf with a hand-operated crane. When the wharf was removed there was a collapse of coffins and skeletons down the canal bank and it had to be reinstated. Close by is the spire of a late medieval red sandstone church. The footway is on the left and into a mural-lined concrete tunnel leading down below the ring road.

At Upper Mitton the canal passes between the Bird in Hand and a graveyard. Although the name Lower Mitton, formerly a Saxon settlement, may be seen on a bridge, the village's name was changed to Stourport-on-Severn when the canal was built. One wharf has a series of arches highlighted with courses of blue bricks. The increase in canalside public houses such as the Rising Sun, Black Swan and Bell Hotel give an indication that the terminus is being reached. York Street Lock is the final lock into the basin. Alongside the lock is a Gothic cottage of 1854 while upstream are interesting wharf buildings that have been restored. This is generally considered Britain's best canal town.

The town was built by the canal company, the only one in Britain built to serve canals, with what would become a 7ha complex of basins at the terminus, in use by 1771. When the Upper

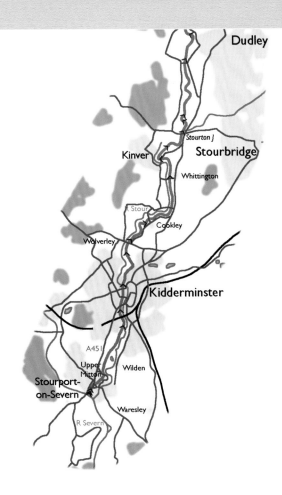

Basin was dug it was only 230mm deep in the centre but deeper at the outsides. In 1956 it was deepened right across. There are Georgian warehouses, one with an overhanging roof on cast-iron columns and one, originally a grain store and now home of Stourport Yacht Club, with a prominent clocktower of 1812 that gives its name to the Clock Basin, also known as the Middle Basin despite being on the upper level 9m above the Severn, clear of floodwater. Two of the three original drydocks in the Lower Basin survive and there are flights of both wide and narrow locks down to the river with a cast-iron footbridge crossing.

Below the inland port is the River Severn, a much older and more major navigation route.

Shropshire Union Canal

The canals that were united to form the Shropshire Union Canal were very different in many respects. The Chester Canal was built between 1772 and 1779 and connected the Dee at Chester with Nantwich. The Wirral Line of the Ellesmere Canal followed in 1793–1797 and was intended to run on past Wrexham to the Severn. They were built to broad gauge to take Mersey flats. A branch was built to Nantwich but the Trent & Mersey Canal refused a connection to the Middlewich Branch for 50 years, afraid that they would lose their northern trade to the Chester route. As a result, the Chester Canal and the Wirral Line became semi-derelict and it was to be the first canal to fail.

The Birmingham & Liverpool Junction Canal was built by Telford, one of his greatest canals, completed in 1835. It was built to narrow gauge as the Birmingham Canal Navigations were narrow, so there was no point in wasting money and water by building wide. In order to challenge the railways it was built with deep cuttings, high embankments, grouped locks and straight lines, so that it was shorter than the Trent & Mersey Canal and had 30 fewer locks. It was the last canal for Telford, who died six months before it opened. Problems caused by two landowners resulted in collapsing embankments, deteriorating health for one of our greatest engineers and technical problems that remain to this day. The link with Birmingham increased through-traffic.

Steam tugs were introduced in 1842 to tow trains of barges but they proved uneconomical and horses were brought back. The Shropshire Union Canal was formed by amalgamation in 1845, the following year becoming the Shropshire Union Railways & Canal Company (although the company also used 'Railway'),

Wolverhampton Boat Club's premises.

with an Act to allow the laying of railways along its canal beds. The canal was leased to the London & North Western Railway and later subsidised by them in order to reach traffic within Great Western Railway and Cambrian Railway territories. The Manchester Ship Canal added more business in 1894 and the route was profitable later than most. Flyboats ran a twice-weekly 30-hour service between Birmingham and Ellesmere Port.

The Shropshire Union Canal runs north-west from Wolverhampton to the River Mersey across New Red Sandstone, through the 30km wide passage of the Cheshire Gate between the Shropshire and Staffordshire hills and then across the Cheshire Plain, with its dairy farming. The canal was needed in an area with few large rivers. It has long straights, sweeping curves, stone-arch bridges and extensive views. It is one of the prettiest canals, rural for 80km to Chester and often tree-lined.

It leaves the Staffordshire & Worcestershire Canal at Autherley Junction in Wolverhampton.

Distance
107km from Autherley Junction to the Manchester Ship Canal

Highlights
Market Drayton Wharf
Bunbury Mill
The old Roman city of Chester
National Waterways Museum at Ellesmere Port, Britian's premier canal museum

Navigation Authority
Canal & River Trust

Canal Society
Shropshire Union Canal Society
www.shropshireunion.org.uk

OS 1:50,000 Sheets
117 Chester & Wrexham
118 Stoke-on-Trent & Macclesfield
127 Stafford & Telford
(139 Birmingham & Wolverhampton)

The ornate Avenue Bridge leading to Chillington Hall.

Telford's Stretton Aqueduct over the A5.

The canal passes under a fine brick towpath bridge and runs through Autherley Stop Lock, which has only a 150mm fall but is intended to protect the Staffordshire & Worcestershire Canal's water at its 94m level. Unusually for a Canal & River Trust canal, gate balance beams are painted grey and white, the canal company colours of 1918–1964, rather than black and white as elsewhere. There used to be a toll office here and there is now a canal shop for provisions. Autherley Boat Club are based in one of a pair of red-and-blue brick horse stables.

The exceptionally tall bridge in Woodseaves Cutting.

There is nothing to indicate crossing the line of the Roman road from Swindon to Water Eaton. The following turnover bridge, accompanied by Wolverhampton Boat Club with its crane, is important not so much because it takes the canal from the West Midlands to Staffordshire but because it moves from a built-up area to open country, a situation that lasts to Chester.

Near Codsall, the canal passes over the young River Penk, fed by Codsall Sewage Works and passing a wood with a nature reserve and trail. The canal has extensive views to the north-west before narrowing into a rocky cutting, a taste of things to come. Upper Hattons bridge has a keystone with the date 1802. This end of the canal was built between 1826 and 1835 but a mason made an error in the 1960s.

Brewood has a landscaped basin where there is the stump of a wharf crane. The canal looks down on the village. Conspicuous is the 13th century red sandstone church of Sts Mary & Chad, its tower with spire and turrets. It has a 16th century font and effigies and 17th century monuments to the Giffords. It also has the grave of Colonel Carless, who helped Charles II escape. This is one of the venues where the king was claimed to have hidden in an oak tree to escape Cromwell's troops. The 18th century Gothic Speedwell Castle, among the Georgian houses in the square, is named after the racehorse that brought the winnings that financed the construction of the castle. The village did also once have a real fort, built on Beacon Hill by the Romans to defend Watling Street. Watling Street was the Roman road that ran from Water Eaton to Wroxeter, now the A5.

The 27km pound heading north begins by passing Wheaton Aston, with the Hartley Arms and a church rebuilt in 1857. The canal crosses the line of the Roman road from Water Eaton to Whitchurch at the start of an airfield that had a squadron of Oxfords and was one of the RAF's biggest training units during the Second World

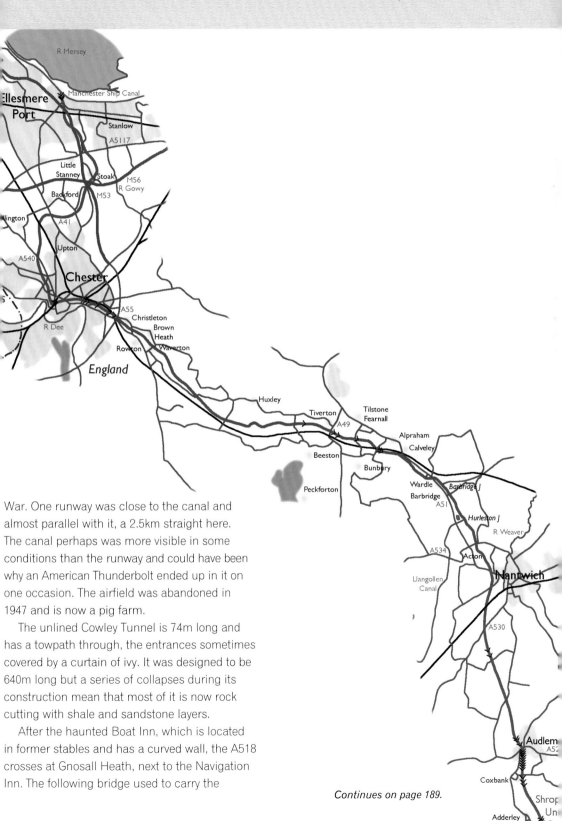

War. One runway was close to the canal and almost parallel with it, a 2.5km straight here. The canal perhaps was more visible in some conditions than the runway and could have been why an American Thunderbolt ended up in it on one occasion. The airfield was abandoned in 1947 and is now a pig farm.

The unlined Cowley Tunnel is 74m long and has a towpath through, the entrances sometimes covered by a curtain of ivy. It was designed to be 640m long but a series of collapses during its construction mean that most of it is now rock cutting with shale and sandstone layers.

After the haunted Boat Inn, which is located in former stables and has a curved wall, the A518 crosses at Gnosall Heath, next to the Navigation Inn. The following bridge used to carry the

Continues on page 189.

Audlem with the Shroppie Fly and Audlem Mill beyond. The railway crane is a recent import.

Stafford to Wellington railway, which was built by the Shropshire Union Railway & Canal Company (SURCC). To the east it passed Gnosall, where the church of St Lawrence has some Norman details, mostly 13th century.

The section through Shelmore Wood should have been simple but it was to prove a problem. Lord Anson of Norbury Park didn't want his pheasant covert disturbed or farmland split up so the 1.6km long Shelmore Embankment had to be built 18m high across lower ground to the west. Construction of the embankment took six years; delaying the opening of the canal by two years.

Norbury Junction was at the end of the Newport Branch of the Shrewsbury & Newport Canal, of which the first 200m remains in water. There are hopes of full restoration. The Canal & River Trust have a maintenance yard here but

the crane is not original. The Junction Inn is a popular public house.

Grub Street Cutting is 1.6km long and 27m deep, not to be confused with Woodseaves Cutting despite the proximity to the first village of Woodseaves. It is crossed at High Bridge by the A519. Telford built a number of very tall arches on this canal and this one is braced across, halfway up. Canals, like railways, became popular routes for lines of telegraph wires. The bracing has the top of a telegraph pole, now without wires, left standing on top, a surreal location for it. A headless horseman sometimes gallops over the bridge. Another peculiarity of the cutting is a black creature like a monkey that has haunted it since a boatman was drowned in the 1800s.

North of the end of the cutting is High Offley with its large 15th century church. The Anchor Inn is an unspoilt boatman's public house. After the arched Shebdon Farm, the Wharf Inn has a crane at the start of Shebdon Great Bank, this one 1.6km long and 18m high, again with a history of having constantly shifted, partly collapsed and consistently leaked, including in 2009.

Knighton has a conservation area around the wharf, which served the 1911 factory where Cadburys received milk from dairy farms by canal, being one of the first users of motorised canal craft. The community has the distinction of having been absolved from paying rates by Charles II, a benefit that was not rescinded until 1990.

For 3km the canal is at the county boundary, but is on the Staffordshire side as if trying to dissociate itself from its name. Beyond Old Springs Hall is a wharf with stone and brick buildings of 1837, which has stone mullioned windows.

The five locks of the Tyrley Flight include two set in a red sandstone cutting, not unique but unusual and attractive with ghostly assistance on the middle lock. Pine trees and gorse flank Peatswood. Opposite is Tyrley Castle. An aqueduct crosses Coal Brook and another passes over the River Tern, both as streams here.

Continues on page 191.

Beginning with the Talbot and including a garden with a clever Bill & Ben feature, the wharf area at Market Drayton is particularly fine, forming the north-eastern extremity of the town. Many of the town's houses are black-and-white half-timbered buildings from the 17th and 18th centuries after a great fire in 1641. It has two gastronomic claims to fame: damsons, and it is said to be the gingerbread capital of the world, using a recipe containing rum.

Robert Clive of India was born here in 1725. Among his exploits were carving his initials in a school desk that remains here, building a dam to flood a shop that refused to pay protection money and climbing the outside of St Mary's church tower and perching on a gargoyle.

The church, built in the 14th century, was founded in the 12th century and has a fine Norman doorway. There has been a street market since 1245 and an annual fair, which used to be a horse fair with dealers coming from Wales.

The town is left with the A53 crossing Lord's Bridge before Tunstall Hall and Victoria Wharf. The canal crosses the route of the North Staffordshire Railway's former line from Market Drayton to Stoke-on-Trent and then is followed by the route of the GWR's line from Wellington to Crewe, the canal having outlived both railways.

After all too short a visit, the Shropshire Union Canal leaves Shropshire for Cheshire, the English county with a greater length of canals than any other. The 15 locks of the Audlem Flight, the Thick, drop the canal 28m.

The A525 crosses at Audlem, where there is a Bridge Inn and old buildings at Audlem Wharf. Audlem Mill is a craft centre and restaurant, originally a 1916 grain and animal foodstuffs mill, with a cantilevered gallery over the canal on the second storey for loading narrowboats. The wharf

crane came from the local railway goods shed. The Shroppie Fly in a former warehouse has the front half of a BCN Joey boat as its bar. The town has timber-framed houses, a Butter Market, a market cross that has become a stone-pillared oak-beamed bus shelter and the Gothic 12th century church of St James the Great, with its well-preserved oak ceiling.

The bottom lock of the flight is by Moss Hall, where a stable is now a private house. There are no more deep cuttings from here to Ellesmere Port, which may or may not be the reason why the plant life in the canal improves. An embankment takes the canal on to the Moss Hill Aqueduct over the fledgling River Weaver. During the Second World War there were plans to make the river navigable from here for 100t barges and to replace the locks with barge lifts so that they could continue up the canal to the West Midlands, but this never happened.

The Wellington to Crewe railway line finally crossed at Coole Pilate. Mickley Hall had a moat but there are slightly more conspicuous defences at Hack House, a 35m radio mast and a radar scanner at what was one of the first radar stations in the Second World War. Mostly below the ground is the 1950s secret nuclear bunker that was prepared for Cold War use, 3,300m^2 of rooms with blast doors, Government headquarters, a decontamination room, a missile early warning system and a BBC studio. On show are nuclear weapons and military vehicles and there is a NAAFI style cafeteria.

Nantwich, a salt town until 1856 because of the underlying rocks, means trading settlement of famous buildings. In 1583 there was a 20-day blaze, following which there was a national collection and Elizabeth I had the town rebuilt with timber from Delamere Forest. The 14th century St Mary's church in red sandstone was greatly restored in 1885 and has an octagonal

Beeston Castle from Wharton's Lock.

tower, fine stone carvings and carved choirstalls, with grotesque oak misericords of knights, monks and dragons and other beasts. Nantwich Town Museum covers the Civil War and fire service and also has a dairy.

Telford had planned to join his line end-on to the Nantwich spur of the Chester Canal on level ground but he reckoned without Dorfold Hall. This is one of the two best Jacobean houses in the county, built in 1616 for Ralph Wilbraham. The driveway and lodge were added in 1862 and there is landscaping by William Nesfield including an avenue of ancient limes and a Spanish chestnut that is said to be a 1,000-year-old relic of Delamere Forest. During the six-week Siege of Nantwich in 1644 it was occupied by each side. The Royalists camped in the grounds. The siege was lifted on January 25th so in remembrance Nantwich people wear holly on Holly Holy Day, the nearest Saturday to this date.

Nantwich Basin, with its marina and restored quality cheese warehouses, was formerly the terminus of the branch from Hurleston. The narrow but direct Birmingham & Liverpool Junction Canal passes through an automatic stop gate and makes an oblique junction with the broad Chester Canal, now a vetch-edged contour canal. The Chester Canal could not afford to pay for the land so the landowners blocked the canal twice, once for a year, until rent was handed over.

The A51 crosses over by Henhull Wharf, used for roadstone. Conspicuous may be a giant sculpture of straw bales, a different design each year by Snugbury's Ice Cream at Park Farm. The demolition of Telford's Junction House here led directly to the formation of the Landmark Trust to restore important old buildings.

Barbridge was briefly a foundry site. It has the Old Barbridge Inn and then the Jolly Tar beyond Barbridge Aqueduct at Barbridge Junction. There was a transhipment dock with a roof spanning the canal where it narrows by a former stop lock. A long low building included a boatmen's mission church with a wooden cross on the end. The blind exit from the Middlewich Branch has resulted in many accidents over the years and these continue today. Much of what traffic has not gone to Llangollen now takes this branch as the Four Counties Ring leaves.

At the very point where the canal traffic leaves, road traffic arrives in the form of the A51 alongside until Calveley, the first time there has been significant traffic alongside on the whole canal, and the Crewe to Birkenhead railway is closing in fast.

Beyond Alpraham, Bunbury Locks are met, a staircase of two in a wonderful heritage setting at Bunbury Wharf, resulting in the intricate Bunbury Shuffle if two descending boats pass two ascending boats in the middle of the flight. The stables have now become steel narrowboat workshops and there is a shop in the lock keeper's cottage, a pump in front, a tiny Edward VII letterbox round the back and a dry dock on the opposite side of the canal. The place has tremendous atmosphere.

In the village of Bunbury is Bunbury Mill, a notable 19th century working watermill, damaged in a storm in 1960 but restored and working. There is also the 14–15th century church of St Boniface.

Tilstone Lock has a lock hut that consists of a single circular room with a bowed wooden front door to maintain the curve of the brickwork. At Tilstone Bank, the bank is used by snowdrops, primroses and bluebells and there is an 1830s mill. The prominent heavily timbered Wild Boar hotel on the south side of the valley is decorated with a range of flags that suggest American tourists might make up a significant part of its customer base.

Wharton's Lock, where the Sandstone Trail footpath crosses, shows Beeston Castle at its best. The castle was built from 1226 by Earl Ranulph of Chester after his return from the 5th Crusade, where he had been inspired by the impregnable hilltop strongholds in Palestine. Built on Beeston Rock, it has a precipice 150m to the Cheshire Plain with the best castle views in England, from where can be seen eight counties, the Welsh mountains, the Pennines and the Wrekin. Richard II was supposed to have hidden a vast treasure in the 110m well but it has not been found. The castle was taken by Simon de Montford in revolt against Henry II, Henry III and Edward I used it to fight the Welsh and it was taken by the Roundheads, followed

Steam Mill in Chester.

by the Cavaliers, in the Civil War, including by eight men who appeared to be part of a large army at night. Royalists held it until six months after the king's defeat in 1645, after which it was destroyed. Earlier, the site had been used by Neolithic farmers, as a Bronze Age hill fort and by Iron Age warriors. The castle has a large outer bailey, D-plan towers, an inner bailey using cliffs on two sides and an early 13th century curtain and gatehouse.

Over the railway, Cheshire Farm makes cheese and one of the widest ranges of dairy ice cream flavours in the country, over 40, including Temptation Tiger and Monkey Poo for Chester Zoo.

The canal sidesteps Rowton Moor where, in 1645, the last major battle of the Civil War was fought. Charles I watched from the city walls at Chester, 4km away, as his soldiers were beaten. The city withstood a two-year siege.

After the Cheshire Cat, a reach of weeping willows leads into Christleton, probably an early Christian settlement. The 14th century almshouses are by John Scott and the 19th century church has a 15th century tower by Butterfield. There is a village green, a pump house and a pond that was a fertilizer marl pit. In the Civil War it was an outpost for the Royalists, who started a serious fire here. Christleton's College of Law is based around an 18th century hall. Riparian owner David Wain played an important role in our inland waterways and set up the canal museum in Llangollen. Back in 1935, a local boatyard may have been the first hirers of canal pleasure craft in Britain. By Quarry Bridge is a 19th century warehouse with a canopy.

There are two Cow Lane Locks, Chemistry Lock and Hoole Lane Lock. Between them is Steam Mill, home to the Artichoke cafe, bar and bistro, and the Old Harker's Arms, trendy city bases in old industrial buildings at Boughton. The Mill Hotel is tucked in beside the low, inner ring road, St Oswalds Way bridge with the Fortress on the opposite corner.

The canal uses the moat at the foot of the city walls. Prominent is the King Charles Tower, from which Charles I watched the battle of Rowton Moor, now a Civil War museum. One of the bridges crossing the red sandstone canyon by Northgate Street is a strange footbridge that reaches between two blank stone walls.

Can anywhere pack so much heritage into so small a space as the centre of Chester? It is based on the Roman street pattern, Deva being named after the Dee, and was a fortress completed in 79 for the 20th Legion by Agricola, the modern name coming from the Old English *ceaster*, a Roman fort. The 3km circuit of the walls is almost complete, the most complete in Britain, being broken only by the council offices. It is one of the best walled Roman cities in Europe and the best surviving Roman wall fortress in northern Europe.

The 1769 Eastgate is on the site of the Roman East Gate. Spud-u-Like in Bridge Street is built over a Roman bath house. The half-excavated amphitheatre, which seats about 7,000 and is the shrine of the Greek goddess Nemesis, is the largest stone amphitheatre in Britain and may have been used for gladiatorial combat. The Roman Gardens have stone and pillars from various sites and a reconstructed hypocaust. The Roman fortress houses the Dewa Roman Experience including a Roman galley. The George & Dragon Inn is on the site of a Roman cemetery and has the ghost of a Roman centurion who passes and returns after 20 minutes, walking through brick walls. Chester has more ghosts than any other English city. After 1,600 years, the changing of the Roman guard now takes place again at noon on summer Sundays.

There were Viking raids in the Dark Ages but the Vikings were driven out by Aethelflaeda by the 10th century and the walls were extended and strengthened. Among the wall's features are the Wishing Steps, which require a person to walk the walls and then run up, down and up again without drawing breath. If they do this they will have their wish come true.

The Saxon minster was rebuilt in the early 10th century as a Benedictine monastery, the most complete medieval monastery complex in the country, parts remaining from 1092. It became the Norman St Werburgh cathedral from 1541 and has 11–18th century architecture of Keuper sandstone. The 1380-carved choirstalls, with the Tree of Jesse and monastic buildings, are the most complete in England, notably the 13th century chapter house, magnificent refectory and cloisters. The massive gateway to the square is the venue for Mystery Plays, the original texts being the most complete in existence, and there is a copy of Handel's marked score for the *Messiah*, first rehearsed here in 1742. There is a 200-year-old cobweb picture, wonderful stained glass and a Renaissance-style font in black marble. A marble bust commemorates Brassey, perhaps the greatest railway contractor of all time. There is also a shrine to St Werburgh, a Mercian princess with miraculous healing powers. In 2002 the cathedral launched Chester Pilgrim Ale made by JW Lees in Manchester. The cathedral is the most popular free-entry destination in the UK.

The market dates from 1139. Katie's Tea Rooms are Britain's largest in one of the oldest buildings, even older than the market. At the time of the Conquest, brewers of poor ale was fined four shillings or ducked in the town pond. At noon on summer Tuesdays to Saturdays there is a town crier at High Cross, where there was bear baiting, stocks, a whipping post and sedan chairs on hire at various times.

Following the Conquest, Hugh Lupus d'Avranches, the cruel nephew of William the Conqueror, was made 1st Earl of Chester and built the first Chester weir to power the Dee mills. Chester Castle of 1069 was built on a Saxon fort

The Northgate lock flight, the railway and the Welsh hills.

site but only the gatehouse remains, the Agricola Tower, dating from the 13th century, now with canal documents. The current building is in Greek Revival style, by Thomas Harrison, and contains the Cheshire Military Museum, with 300 years of Cheshire regiments plus Ypres trenches and the Victoria Cross of Todger Jones, who captured 120 Germans single handed.

The Three Old Arches of 1200 is England's oldest shopfront. In a city of many half-timbered black-and-white buildings, the Rows are exceptional. These unique 13th century shop galleries on two levels through half-timbered Tudor buildings are like being on a galleon.

The expression 'There's more than one yew-bow in Chester' made to girls who get jilted is a reference to the large number of local archers lost at Agincourt, Crécy and Poitiers – Chester archers being the best in the country. Not all girls were losers, however. The story goes that a medieval mayor attempted to force his daughter to marry Luke de Taney. In a game with him she hit the ball over a wall, sent him to fetch it, escaped through the Pepper-gate while he was gone and eloped with a waiting Welsh knight. The gate has been locked ever since. Because there were so many border brawls, Henry IV banned all Welshmen from carrying any weapons

Britain's largest surviving canal dockyard, sadly dilapidated but with scope for restoration.

except eating knives and instituted a sunset curfew for them.

Bishop Lloyd's house of 400 years ago has a richly carved front with biblical scenes and animals. Chester Heritage Centre is in the old St Michael's church. St Mary's centre has exhibitions in the 14/15th century church of St

Taylor's boatyard with the main line on the left, and the link to the River Dee on the right.

Mary, an excellent example of Perpendicular styling with a Tudor nave roof, two 17th century effigies, medieval stained glass and a wall painting at the top of one of England's steepest streets.

Chester Visitor Centre, the biggest in Britain, is in a Grade II building with a history of Chester, including the Rows and a recreated 1850s Victorian street. The Grosvenor Museum has exhibits from Roman times to the present day, including a Roman graveyard, a period house, a Victorian schoolroom, Chester silver, Anglo-Saxon coins and natural history. The Chester Toy & Doll Museum has 5,000 items from 1830 onwards, including the biggest collection of Matchbox toys in Europe. There is a Gothic-style Victorian town hall with its 49m tower containing the Chester Tapestry. The Victorian Eastgate jubilee clock of 1897 is claimed to be the second-most photographed clock in the world after the Houses of Parliament.

There are festivals for summer music, jazz, folk, literature, fringe, street processions and cheese rolling, an international horse show since 2002 and an international church music festival begun in 2003.

The three Northgate Locks were hewn out of solid sandstone. The Chester to Holyhead railway passes low over the flight so that boaters look down on to the tracks from above, with a background of Welsh hills, the Welsh border being less than 2km away at this point. It was the collapse of Stevenson's nearby wrought-iron-braced, cast-iron railway span over the Dee in 1847 that hastened the use of straightforward wrought-iron then steel-plate girders. The Dee is 200m ahead as the canal turns sharply right under a road bridge with some complex girders supporting a tight corner. The connection, straight ahead originally with two further locks, goes through a hairpin arrangement that adds 500m to the journey down through three locks.

In the Middle Ages, Chester was the most important port in northern England, exporting cheese, candles and salt, but it gradually silted up. Although part of the Roman harbour wall remains, the Roodee is now part of the site of Chester races, Britain's oldest horse races, running since 1539. The Water Tower was a port defence structure, built in 1322.

The Old Dee Bridge, built in 1387 by Henry de Snelleston, has had various rebuilds but was

the only bridge across the Dee in Chester until the 19th century. A Gothic-style hydroelectric power station was added at the northern end in 1913, the first for an English city, converted to a pumping station in 1951. Thomas Harrison designed the Grosvenor Bridge in 1802 but delays meant it was not opened by Princess Victoria until 1832. At 61m it was the longest stone-arch bridge in the world. It remains the longest in Britain and the fourth-longest in the world. Chester weir is Britain's oldest surviving mill dam and is where an Environment Agency report first recorded that canoes do not disturb fish. Rowers have Chester Regatta, the world's oldest, which has been running since 1733.

Telford built two magnificent warehouses on the canal; sadly both were burnt down in recent years by arsonists. The one in Chester has been rebuilt as the Telford's Warehouse public house and entertainments venue. By the junction is Tower Wharf, filled in during the 1950s but re-excavated in 2000. A development of housing, offices and restaurants has taken place around it. Harvest House was the canal company headquarters and is still the maintenance centre, with a Georgian office, small warehouse, manual crane and neat cast-iron roving bridge. By the towpath is a plaque to Tom Rolt, the local man who used his Shropshire Union boat *Cressy* to inspire leisure use of the canals and who is the main figure responsible for the canal's§ revival. Taylor's Boatyard is a dry dock in the fork with astable block, forge, former steam sawmill and workshops. The dilapidated Shropshire Union dockyard is on the west side of the junction – the largest surviving. The restoration potential here is enormous. A century ago there were 448 houseboats on the canal in Chester. It is a very different canal today.

The Deva Aqueduct is crossed just before the A5480 crosses over, followed by the former Mickle Trafford to Connah's Quay railway bridge at Abbot's Meads. After Blacon with its extensive housing estates there is open country, partially wooded, with swallows but hardly a building in sight for 6km, Chester being completely hidden. The most garish intrusions are cyclist signposts that look like gaudy sculptures. There are powerlines and the A540 crosses. The Crewe to Birkenhead railway with Merseyrail, on a splendid 1839 red sandstone viaduct with 11 skewed arches, makes its final appearance between Moston and Mollington. The view north-west between the railway and the A41 is unbelievably rural. Somebody picked a prime spot for the county offices at Backford.

Even Chester Zoo, up Butler Hill at Upton Heath, is virtually invisible. The UK's largest and most popular zoo, it is one of the finest in the world, carrying out much conservation work with endangered species. Covering over 50ha, it has more than 520 species, over 7,000 animals including jaguars, black rhinos, red pandas, marmots, Britain's only komodo dragons and birds of paradise, a bat cave, the world's largest elephant house, buffy-headed capuchin monkeys and macaws in a forest close to the canal. They have the biggest monkey house in Europe, the largest social group of chimpanzees in the UK, orchids, 80,000 plants including Roman and South American gardens, a monorail and a waterbus running on mini canals. A new marina development is to be built next to the zoo and a huge expansion is to include rainforest biodomes with a canopy walkway and water ride, making it the largest animal attraction in Europe by 2018.

Little Stanney has the Blue Planet Aquarium, the UK's largest with over 2,500 fish. This is home to more than a score of sand tiger sharks in Europe's largest collection, the world's longest moving underwater walkway and Europe's largest indoor rockpools. Close by, at Wolverham, is the McArthur Glen Cheshire Oaks designer outlet village, Europe's largest with 140 stores. 2003's local tourist board guide said it has 'Children's play areas, easy access to motorways.' There is also the Iwerks Extreme Screen Attraction 3D cinema, the first in Britain. The Coliseum is a 15-screen multiplex cinema with Megabowl, two nightclubs, restaurants and the UK's largest single-storey book/music/film store.

The M53 crosses back at Junction 9 and ahead lie Ellesmere Port docks of 1833. The name was changed from Netherpool by the Ellesmere Canal Company in 1796 to be the port for the Shropshire town with which they hoped to connect directly. A transhipment port, it is the finest canal port in England and is now the home of the National Waterways Museum; Ellesmere Port; and Britain's premier canal museum with the world's largest collection of inland waterways craft – over 60 of them – including a 1912 tunnel tug and an Iron Age dugout. Other exhibits include dock workers' cottages of the 1840s and 1950s, a power hall and working pumphouse with steam, gas and diesel engines, painted boat ware, tools, plans, documents, stables, a

working forge and a cafe. It was opened in 1976. A popular attraction, the museum is visited by many groups of children during term time. An entry fee is likely to be requested if following the canal through the site.

There are basins to the left of the lock flight and then the canal turns right past a lighthouse of 1795. This was to guide shipping in from the River Mersey although, even at high water, that is at least 500m away across marsh now and the Manchester Ship Canal directly fronts the Shropshire Union Canal. It is closed to small boats and even narrowboats have to raft up in pairs. Nearby are a Ro-Ro ferry terminal, a container terminal and large ocean tankers and other craft passing down the Manchester Ship Canal. For many, over the decades, this point has not marked the end of the journey so much as the start of something much bigger.

Telford's lighthouse by the Manchester Ship Canal.

27 Llangollen Canal

The names 'Llangollen' and 'Canal' do no more than reflect the confused history of a waterway that bears little resemblance to what was originally planned. Built by Telford and Jessop, it was opened below Pont Cysyllte in 1805 and from Llangollen three years later, not that Llangollen featured in the original plan. It was to be a canal running north–south from Chester via Wrexham, Chirk, Ellesmere, Frankton and Weston to Shrewsbury and the River Severn, taking coal from Ruabon and bringing lime to numerous kilns. A change was made to a cheaper east–west link via Whitchurch to the Chester Canal, to become the main line of the Shropshire Union Canal. The plan to link the Mersey, Dee and Severn was not to happen as a direct line. Had the change been made sooner, the section above Frankton would not have been built and the feeder from the Dee would have been along the south side of the river. However, the magnificent Pontcysyllte Aqueduct was complete, leading up to what was intended to be a 4.2km tunnel to the north of Trevor. Cutting of this never began.

To supply the feeder, Telford built Horseshoe Falls, the falls actually being J-shaped, despite the name. 140m long and 1.2m high, of masonry with a cast-iron capping, it provides a reliable source of 5,500m³/day from the River Dee, supplying water for domestic use at the far end of the canal at Hurleston Junction and canal water as far apart as the northern end of the Montgomeryshire Canal and the Manchester Ship Canal. A dam increased the depth of Llyn Tegid at Bala to enhance supplies.

Repaired section clinging to the hillside in Llangollen.

Above the weir is Llantysilio Hall. Parts of the interior of the Victorian church at Llantysilio come from Valle Crucis Abbey. The glass in the south window is 16th century.

Below the Horseshoe Pass is Valle Crucis Abbey. The Cistercian abbey was founded in 1201 and fell into disrepair after the Dissolution of 1535, when it was the second-richest Cistercian monastery after Tintern. It has been partially restored. It has sculpted memorial slabs and includes parts of the church, 14th century chapter house and cloister. It has a notable west front with carved doorway and rose window. The chapter house has a rib-vaulted roof near the original fishpond, the only monastic one in Wales. The name comes from the valley of the cross. The 2.4m inscribed stone cross, just up the valley, was probably 9th century and known as Eliseg's Pillar after the Prince of Powys who built Castell Dinas Bran and was killed in battle in 603. The top third of the cross has been lost, having been pulled down in the Civil War. Owen Glyndwr met the abbot one morning and remarked that he was up early. The abbot replied that it was Glyndwr who was early, by a century, perhaps a reference to the rise of the Tudors in 1485, to whom he was related. Glyndwr was not seen in public again after this meeting.

A field that has been the home of the International Musical Eisteddfod since 1947 now has a permanent pavilion. The Eisteddfod is an international gathering of music, song, dance and poetry each July. The inaugural National Eisteddfod was also held in Llangollen and returns here from time to time but it changes its venue around Wales every year.

Approximately fifty thousand people per season take horse-drawn boat trips from Llangollen Wharf, the longest-running trip boats on the entire canal system, having been established in 1884 and more recently operated in three boats that are pulled by Welsh cobs. The wharf building is now a restaurant.

Distance
75km from Llantysilio to Hurlston Junction
Highlights
Castel Dinas Bran
Pontcysyllte Aqueduct, the longest and tallest
Chirk aqueduct and viaduct in Britain
Navigation Authority
Canal & River Trust
OS 1:50,000 Sheets
117 Chester & Wrexham
118 Stoke-on-Trent & Macclesfield
125 Bala & Lake Vyrnwy
126 Shrewsbury & Oswestry

The town is called after Collen ap Gwynnawg ap Clydawg ap Cowdra ap Caradog Freichfras ap Llyr Merim ap Yrth ap Cunedda Wledig. St Collen's church has a carved oak roof which may have benefited in the 13th century from the skills of craftsmen working on Valle Crucis Abbey, the church being enlarged in 1865. Also enlarged was one of the Three Jewels of Wales, the bridge crossing the Dee below Llangollen Town Fall. Begun as a stone packhorse bridge in 1282, it was rebuilt in sandstone in its present style about 1500 by John Trevor, Bishop of St Asaph. It was being lengthened to accommodate the railway in 1865 and widened to 6.1m in 1873 and 11m in 1969, its four arches retaining their pedestrian refuges over the cutwaters.

A large school is left on the uphill side as the canal moves away on a section rebuilt with a heavy concrete lining in 1985. This section is narrow and has a long line of moored narrowboats in the summer. Across the river is the Hand Hotel, claimed by Kilvert in the 1870s to be the only hotel in Wales with a Welsh harp (as distinct from the easier-to-play English harp). Gale's Wine Bar, along the road, was opened in the mid 1970s, the local paper protesting it was not what Llangollen wanted as it would attract Hell's Angels and the like. With old church

The canal begins cut into an ivy-clad rock wall above the Serpent's Tail.

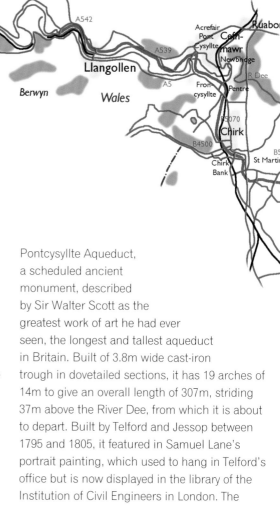

pews for seating and a harpist sometimes playing in the evening, it has done so much to improve the quality of the town that it has spawned a row of look-alikes.

The canal breaks clear of its rock cutting to reveal steep sheep pasture leading up to the 340m peak, topped by Castell Dinas Bran, the castle of the crow, the legendary castle of the Holy Grail. The home of Gryffudd ap Madoc, son of the founder of Valle Crucis Abbey in the 12–13th centuries, it was 88m x 43m and nearly impregnable but was burned in 1277 by the Welsh defenders in the war between Llewellyn ap Gryffudd and Edward I. On a clear day it has views of the peaks of Snowdonia. Behind is the grey carboniferous limestone escarpment of Creigiau Eglwyseg, packed with fossils.

An old lattice footbridge carries the Offa's Dyke Path over the canal. While the path was opened in 1971, the dyke was probably built in the 780s to keep the Welsh at bay.

Narrowboat moorings at Trevor are sited in the remains of the Ruabon Branch on what was to have been the main line from Chester. Trevor Wharf is served by the Telford Inn.

The other way, though, is one of the seven wonders of the canal world, the Grade I

Pontcysyllte Aqueduct, a scheduled ancient monument, described by Sir Walter Scott as the greatest work of art he had ever seen, the longest and tallest aqueduct in Britain. Built of 3.8m wide cast-iron trough in dovetailed sections, it has 19 arches of 14m to give an overall length of 307m, striding 37m above the River Dee, from which it is about to depart. Built by Telford and Jessop between 1795 and 1805, it featured in Samuel Lane's portrait painting, which used to hang in Telford's office but is now displayed in the library of the Institution of Civil Engineers in London. The masonry piers are partly hollow, built with a mortar of ox blood, water and lime. The trough was made waterproof with Welsh flannel dipped in boiling sugar. The towpath is built over the trough so that boats do not reach to both sides and water can flow past them, rather than a boat acting as a piston and pushing all the water across the aqueduct. The project brought together a construction team later to work on other major projects such as the Menai Bridge and the Caledonian Canal. The flimsy handrail is of slight consolation for people with vertigo but on the west side the only protection is an upstand of some 300mm. The canal is dewatered periodically with a dramatic discharge from the centre of the aqueduct to the river far below.

Upstream is Pont Cysyllte itself, a three-span sandstone-arched bridge of 1696 over the river.

A cutting hides Chirk Castle. Built in 1284–1310 by Roger Mortimer for Edward I, it was bought in 1595 by Sir Thomas Myddleton, whose family have owned it ever since. It is the only border castle to have been occupied continuously since it was built. Restored, it has a 30m Long Gallery of 1678, a Great Saloon,

a deep dungeon, walls up to 4.6m thick round a central courtyard, large round corner towers and D-plan intermediary towers, all set in 1719 parkland and formal gardens by the Davies brothers, behind magnificent wrought-iron gates. There are elegant state rooms with fine ceilings and walls, Chippendale furniture, James I tapestries, an 18th century staircase and a ghost. The family arms include a red hand. This is said to represent former misdeeds and that this may be removed when a prisoner can survive in the dungeon for ten years. One prisoner almost managed it.

The 420m long Chirk Tunnel was built by Telford in 1801, with a slightly crooked bore and flared ends, which make the bore appear larger than it really is. All three tunnels on this canal have towpaths, unusual for the time, avoiding legging. This is the longest British tunnel regularly used by walkers.

The river marks the border between the Welsh county of Wrexham and the English county of Shropshire. This is Marches country. The Bridge Inn was important as the first public house in England in the days when Sunday drinking was not allowed on the Welsh side of the border.

Despite the terrain, there are only two locks in the 51km between Llantysilio and Whitchurch. Ironically, they come in the much more gentle country at New Marton, a village marked by a wind pump. These are the busiest locks in Britain with 9,500 movements per year.

The Jack Mytton Inn is a prominent canalside building at Hindford. Just after the Oswestry to Whitchurch railway crosses, a neat brick roving bridge takes the towpath across before the Narrow Boat Inn.

There are extensive views southwards over the Perry valley, for the next 3km approaching Tetchill, to hills south of the River Severn. The Montgomery Canal line of the Shropshire Union Canal joins at Lower Frankton. Following the contours, the canal makes a near 180° bend before Tetchill.

Ellesmere, a town with Saxon origins, begins with Ellesmere Depot and Yard, joinery shops and forge, now including interesting private dwellings. The boat repair shop is still in use and the Canal & River Trust provide services for powered craft. On one occasion a floor collapsed under British Waterways senior management ironically having a health and safety meeting. There is a picnic area as the Ellesmere Branch goes off to the town centre, the canal's eastern terminus initially. The motte and bailey castle site is now a bowling green in this 18th century market town with its Georgian houses. St Mary's church has medieval origins, a medieval chest made from

a solid oak block, a 15th century octagonal font and many fine effigies.

The meres are kettle holes surrounded by kame moraines. Blocks of ice buried in the ground at the end of the Ice Age 15,000 years ago gradually melted without any drainage paths leading away. Local legend says that there was just one well in a field. A new tenant refused to let the villagers use it or they were charged by the bucket. They prayed for assistance and the well overflowed so that there was free water for all but the tenant still had to pay the rent for his now flooded land. Another legend has 16th century highwayman Sir Humphrey Kynaston landing on his horse after a 14km jump from Nesscliffe, perhaps assisted by the belief that the horse may have been the Devil. He did make prodigious but more plausible jumps on horseback, including down a cliff from his Nescliffe cave to escape. These days the meres are dark pools surrounded by trees. The Mere, at 45ha, is the largest. The canal skirts Blake Mere and Cole Mere. The latter has the 28ha Colemere Country Park, wildfowl and information on the geology, flora and fauna of the meres and mosses and the remains of limekilns in this Site of Special Scientic Interest.

For 1.6km the canal runs straight and wide, crossing back into England to a canal junction dominated by a red-brick house with circular insets. The Prees Branch never reached Prees but it did get to Edstaston, hence its alternative name of Edstaston Branch. These days, it only reaches Waterloo. Whixall Moss roving bridge is a wooden structure with lattice sides, adding to the character of the area.

Platt Lane has a typical slated, red-brick wharf and warehouse. There are lime kilns in a garden at Lower Tilstock Park and the site of a brickworks on the right before the former Cambrian Railway line from Oswestry to Whitchurch crossed over. These local industries can be easily missed, unlike

Castell Dinas Bran looks down over the autumn colours along the canal.

the spectacular views to the left over Wales, the border of which is close to the left bank of the canal for the last time.

While this section has been lock-free, the same cannot be said of access drawbridges that lower almost to water level. Three come in a kilometre on the approach to Whitchurch.

The partially restored Whitchurch Branch leaves at Chemistry. Houses have been built on the old line of the canal but it is hoped to extend the branch. Much of the town remains hidden from the main line. Whitchurch was the site of a Roman fort, Roman artefacts being shown in the local museum. The original Norman church of St Alkmund collapsed in 1711 and was rebuilt two years later. The town is home to the Joyces, the oldest tower clock-makers in the world, eight generations having run the firm for over 300 years; their work includes the church clock of 1849. A market town, it is claimed as the home of Cheshire cheese.

While the canal doubles back round the head of a steep valley, the A41 strides straight across on an embankment and crosses over the canal

again. It makes a final pass over at Grindley Brook, this time on the line of the Roman road from London to Chester.

Since Horseshoe Falls there have been only two locks but that now changes. Grindley Brook Upper Lock is a staircase of three accompanied by a bow-fronted cottage, the Lockside Stores and @29 Cafe. Grindley Brook Lower Lock is made up of three separate locks after the road. The Horse & Jockey public house stands back from the canal. In 2009 there was a breach after a culvert collapsed.

After the locks, the canal is crossed by a massive former railway embankment using a blue-brick arch with magnificent skewed brickwork. Here the canal meets the Cheshire border and this is a point from which various long-distance footpaths strike out: the Shropshire Way, the Cheshire Way and the Sandstone Trail to Frodsham.

The canal now runs along the bottom of a deep valley with Hinton Hall prominent on the east

Two magnificent structures, Chirk Aqueduct and the adjacent railway viaduct, crossing the Ceiriog valley.

Pontcysyllte Aqueduct soars above the treetops with minimal protection from the elements.

ridge and the slope rising on the other side from Land of Canaan to Bell o' the Hill Farm. Hinton Bank Farm produces blue Cheshire cheeses.

The Willeymoor Lock Tavern stands next to its namesake. In 1983 the pub designed and had built a metal bridge over the canal to give access to the pub and to be used by walkers and emergency services rather than having to walk across the lock gates. A very sensible decision as lock gates are not the easiest of moves even before starting drinking.

A lifting bridge at Wrenbury dates from 1982 and tilts to 80°, steeper than its listed predecessor, which caused accidents. There are two mill buildings, the older of which dates

from the 16th century. The miller had his own fleet of narrowboats. The Dusty Miller by the canal is accompanied by the Cotton Arms down the road towards the battlemented church of St Margaret with its 16th century west tower, 18th century chancel and pulpit and fine 19th century monuments.

The four Hurlston Locks are fitted with experimental low-geared enclosed paddles introduced in the 1970s. The penultimate lock has been suffering with a wall moving in, making it too narrow for some boats to pass, otherwise the canal would be even busier.

At the foot of the flight the Llangollen Canal meets the main line of the Shropshire Union Canal at Hurlston Junction. Here it is just 2km from the end of the Middlewich Branch.

28 Shropshire Union Canal: Middlewich Branch

The Middlewich Branch of the Shropshire Union Canal was opened in 1833 by the Chester Canal Company and engineered by Thomas Telford as a short but important link between the Shropshire Union Canal's main line and the Trent & Mersey Canal. It provided a faster and less locked route than one via the Staffordshire & Worcestershire and Trent & Mersey Canals. In 1888, there was an unsuccessful attempt to use narrow-gauge railway steam engines to tow boats.

It leaves the main line at Wardle. Barbridge Junction, the branch point, is only 2km from Hurlestone Junction where the Llangollen Canal joins the main line on the opposite side and so traffic from Llangollen and Chester both feed through to the Trent & Mersey on the Middlewich Branch. It is part of the Four Counties Ring.

The line is almost entirely rural, reaching across the gently rolling Cheshire countryside, more often than not through meadows with black-and-white Fresian dairy cattle. Studded with oak trees, it is a pleasantly pastoral scene that contrasts with the activity on the canal.

Cholmondeston Lock, like other locks on the canal, is built of heavy red sandstone blocks. The locks are deep and narrow, daunting for narrowboat users, many of whom are novices emerging from Venetian Marina, one of the largest canal marinas in England, Venetian Cafe and a caravan park, next to where the Chester to Crewe railway line crosses the canal.

The Middlewich Branch leads away from the main line.

Looking over the Top Flash and River Weaver to Winsford.

Oak trees on each side of the canal drop away as it runs out on to an embankment, carrying it high across the Weaver valley, crossing the river itself on a tall aqueduct, which is inconspicuous from the canal.

Many of the bridges have been painted matt grey at some stage in their past, now peeling to reveal the original red brickwork. A bridge above Church Minshull shows how the canal is clinging to the side of the valley. The village itself lies way below, a community of timber-framed houses powered by electricity from the miller's millwheel until 1960. Aqueduct Marina has been created here, with the Galley Cafe

The canal curves on past a noticeable aerial by a wood. A stable block, formerly used to change flyboat horses, is now converted to residential use beside the canal. There are various large farms on the right bank until the canal is above the Top Flash. The view from here across the River Weaver far below and down its valley to Winsford and the start of the Weaver Navigation is phenomenal. To get to the river by water, though, involves a detour around Northwich as far as the Anderton Boat Lift.

For the first time, the canal comes into a built-up area as it passes recent housing in Middlewich, complete with a private landing stage serving all the gardens. The canal moves under several weeping willows and bridges, the last being close to the line of the Roman road that approached this important salt mining town from the south.

Wardle Lock was named after the fact that it is at the west end of the Wardle Lock Branch. The branch, a mere 100m long, was opened in 1829 mainly as a convenient place to intercept and extract tolls from barges travelling from Chester on Shropshire Union Canal waters before they could get on to the Trent & Mersey Canal to reach wharves in Middlewich.

Distance
16km from Barbridge Junction to the Wardle Lock Branch
Highlights
View from the Weaver Aqueduct
The 100m Wardle Lock Branch
Navigation Authority
Canal & River Trust
OS 1:50,000 Sheet
118 Stoke-on-Trent & Macclesfield

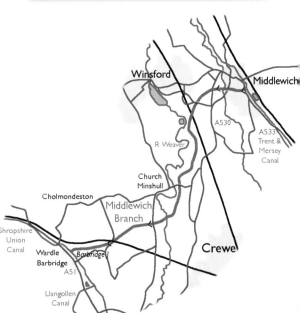

29 Trent & Mersey Canal

Promoted by the Liverpool corporation and the Staffordshire pottery owners under Josiah Wedgwood, the Trent & Mersey Canal was formerly the Grand Trunk Canal that linked the ports of Liverpool and Hull. It was part of the Grand Cross that joined the Mersey, Trent, Severn and Thames. Built between 1766 and 1777, it was James Brindley's most ambitious project and was one of the most successful canals. It was later bought by the London, Midland & Scottish Railway, though regular freight traffic stopped in the late 1960s.

Despite the name, the canal never actually comes within sight of the Mersey but follows the Trent for the greater part of its length. It originally joined the Preston Brook Branch of the Bridgewater Canal at the northern end of Preston

The Anderton Boat Lift links the Trent & Mersey Canal and the Weaver.

Brook Tunnel. This was subsequently extended to 1.133km, giving the unique situation of canals joining inside a tunnel. This was the first major tunnel to be built and is the ninth-longest canal tunnel still in use.

A fairly straight path leads over Preston Brook Tunnel, past a series of vent shafts and the Tunnel Top public house.

Dutton Stop Lock has a 150mm rise to isolate from the Bridgewater the water of this long pound, which runs from Middlewich. By Dutton Hall, the canal is 15m above the River Weaver and Weaver Navigation, looking down on Dutton Locks and to the Dutton Arches of 1836, twenty 18m spans carrying the West Coast Main Line 20m above the river.

At 129m and 174m long, Saltersford and Barnton tunnels follow in quick succession with a wide reach between them. Tunnelling methods were in their infancy when they were cut and they are not particularly straight, although at least a segment of light is visible right through. Tugs were fitted with side wheels to act as fenders.

British Waterways' most difficult rebuild took place opposite the Stanley Arms. The Anderton Boat Lift was reopened in 2001 after an 18-year closure. This is one of the wonders of the canal world, the world's first boat lift, and is unique in Britain although it was later emulated in Belgium, France, Germany and Canada. When the Trent & Mersey was opened, chutes down to the Weaver brought much new trade. In 1875 Sir Edward Leader-Williams erected the steel viaduct and pair of guillotine gates to allow boats to be transferred the 15m between the two waterways as salt had previously been transferred by handcart from narrowboats. The boat lift's 22.9 x 4.7m tanks weighed 91 tonnes empty and 252 tonnes full and were counterbalanced. This was the most daring-ever use of hydraulics, water from a cylinder below the descending tank being used to force up the other tank, assisted by limited steam power for the eight-minute cycle. In 1903 the locks were powered by electricity and allowed to operate separately, becoming individually counterbalanced in 1908. The hydraulics were replaced by thirty-six 1.8m diameter wheels for each tank, each loaded by a 7t weight stack, the cycle time now being reduced to a mere six minutes.

Distance
150km from the Preston Brook Branch to Derwent Mouth

Highlights
Anderton Boat Lift, the world's first boat lift
Lion Salt Works Museum
Harecastle Tunnels
Wedgwood visitor centre
Staffordshire County Museum at Shugborough Park, a Victorian-era working estate
Fradley Junction
Shadlow Heritage Centre and Wharf

Navigation Authority
Canal & River Trust

Canal Society
Trent & Mersey Canal Society Ltd
www.trentandmerseycanalsociety.co.uk

OS 1:50,000 Sheets
117 Chester & Wrexham
118 Stoke-on-Trent & Macclesfield
127 Stafford & Telford
128 Derby & Burton upon Trent
129 Nottingham & Loughborough

Looking south from Barnton tunnel towards the A533 crossing.

The Anderton Nature Park has been opened in the area adjacent to the lift with its flashes and has recorded over 220 species of bird, 25 species of butterfly and nearly a score of dragonfly and damselfly species, including several rare migrants. The last week in July usually offers the chance of seeing the maximum number of species.

Ingram's Lion Salt Works is now a museum, Britain's last works to produce salt by evaporating brine, operating from 1842 to 1986. It has a horizontal steam engine, nodding donkey brine pump and 1900s pitched-roof railway salt wagon, but the most interesting feature is right next to the canal where the arches at the tops of the boiler room doorways are now at towpath level, such has been the settlement.

Middlewich Bottom Lock with a formerly split footbridge.

The canal arrives in Middlewich, the Roman Salinae, the middle saltworks town between Northwich and Nantwich, *wic* being Old English for working place or *wych* the Anglo-Saxon for salt. Middlewich Big Lock gives access to the final broad-beam pound. The lock gate was operated winch and chain because of the proximity of a bridge. The adjacent Big Lock public house has its name carved in a brick scroll over the door and on a fine day it is necessary for towpath users to walk between tables of drinkers. The Folk & Boat Festival takes place in June. Saltworks here included those of Murgatroyd's and Seddon's. There is a moat up the hill.

After Middlewich Top Lock, the short Wardle Canal leads off to connect with the Middlewich Branch of the Shropshire Union Canal. The Trent & Mersey Canal now becomes part of the Four Counties Ring to Great Haywood.

Wheelock may take its name from the Welsh *chwel*, to turn, because of the winding river. The A534, formerly the Sandbach to Nantwich turnpike, crosses before the Commercial Hotel. Opposite the wharf was a fustian mill, which was to become a phosphoric acid works during the First World War and then a printing works.

So far, the canal has been remarkably lock-free but that now changes. The two Wheelock Locks start Heartbreak Hill to Kidsgrove, 26 locks in 11km, often duplicated to remove bottlenecks.

Lawton was Saxon for hill farm. All Saints' church at Church Lawton dates from 1180, on a Saxon mound, and has a Norman doorway, 1540 spire and red-brick nave of 1803, following a fire four years earlier. The remains of St Werburgh rested here in 707 on the way to Chester for burial. Lawton Hall looks Gothic but is only 17th century.

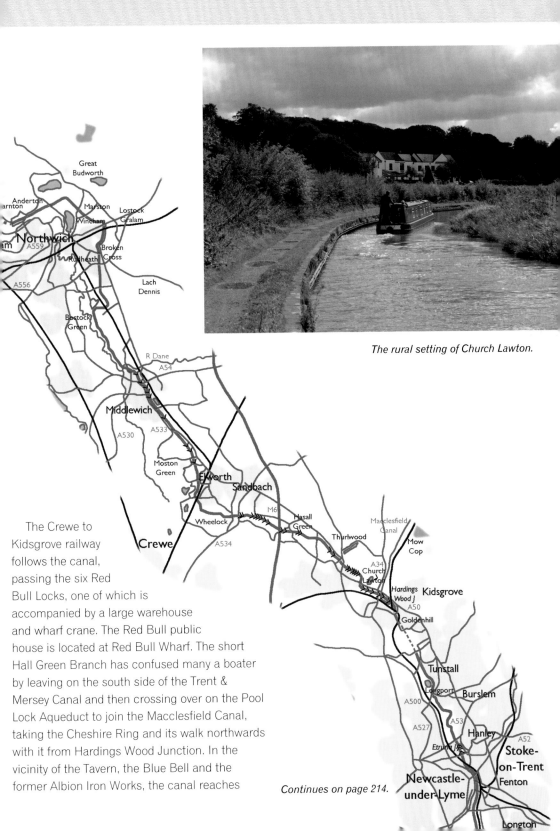

The rural setting of Church Lawton.

The Crewe to Kidsgrove railway follows the canal, passing the six Red Bull Locks, one of which is accompanied by a large warehouse and wharf crane. The Red Bull public house is located at Red Bull Wharf. The short Hall Green Branch has confused many a boater by leaving on the south side of the Trent & Mersey Canal and then crossing over on the Pool Lock Aqueduct to join the Macclesfield Canal, taking the Cheshire Ring and its walk northwards with it from Hardings Wood Junction. In the vicinity of the Tavern, the Blue Bell and the former Albion Iron Works, the canal reaches

Continues on page 214.

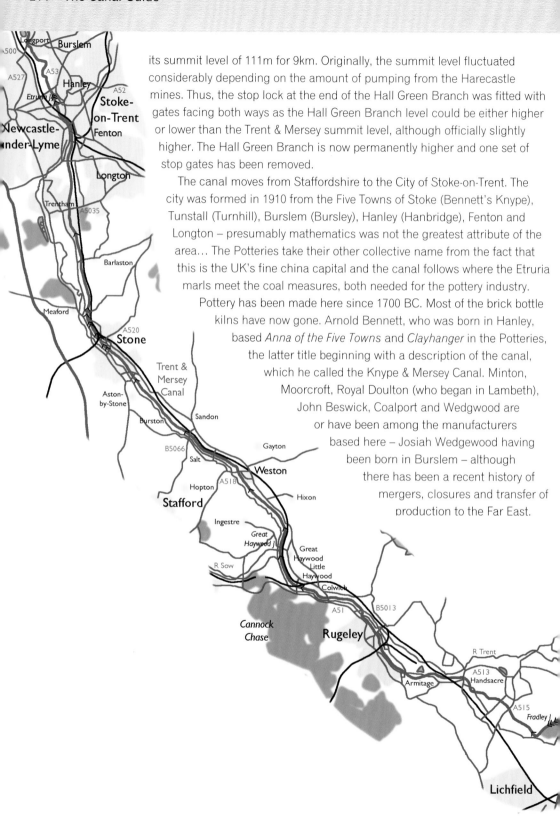

its summit level of 111m for 9km. Originally, the summit level fluctuated considerably depending on the amount of pumping from the Harecastle mines. Thus, the stop lock at the end of the Hall Green Branch was fitted with gates facing both ways as the Hall Green Branch level could be either higher or lower than the Trent & Mersey summit level, although officially slightly higher. The Hall Green Branch is now permanently higher and one set of stop gates has been removed.

The canal moves from Staffordshire to the City of Stoke-on-Trent. The city was formed in 1910 from the Five Towns of Stoke (Bennett's Knype), Tunstall (Turnhill), Burslem (Bursley), Hanley (Hanbridge), Fenton and Longton – presumably mathematics was not the greatest attribute of the area… The Potteries take their other collective name from the fact that this is the UK's fine china capital and the canal follows where the Etruria marls meet the coal measures, both needed for the pottery industry.

Pottery has been made here since 1700 BC. Most of the brick bottle kilns have now gone. Arnold Bennett, who was born in Hanley, based *Anna of the Five Towns* and *Clayhanger* in the Potteries, the latter title beginning with a description of the canal, which he called the Knype & Mersey Canal. Minton, Moorcroft, Royal Doulton (who began in Lambeth), John Beswick, Coalport and Wedgwood are or have been among the manufacturers based here – Josiah Wedgewood having been born in Burslem – although there has been a recent history of mergers, closures and transfer of production to the Far East.

The northern end of Telford's new bore for Hardcastle Tunnel with Brindley's original bore to the right.

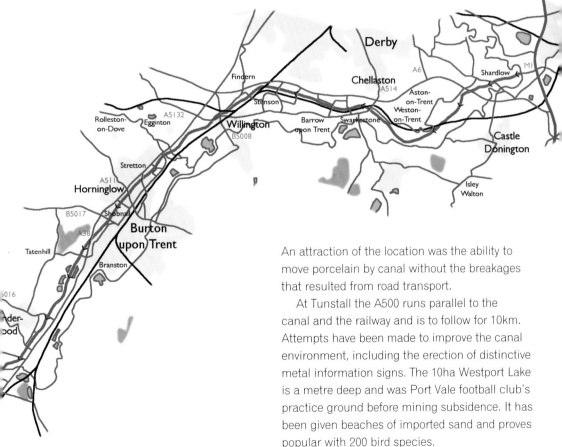

An attraction of the location was the ability to move porcelain by canal without the breakages that resulted from road transport.

At Tunstall the A500 runs parallel to the canal and the railway and is to follow for 10km. Attempts have been made to improve the canal environment, including the erection of distinctive metal information signs. The 10ha Westport Lake is a metre deep and was Port Vale football club's practice ground before mining subsidence. It has been given beaches of imported sand and proves popular with 200 bird species.

At Longport (Bennett's Shawport) is the Grade II Georgian Duke of Bridgewater public house, someone not directly involved with this canal. A bottle kiln remains among the buildings of the Middleport Pottery, where Burleigh's factory is open for tours. Davenport was set up here in 1793.

At Burslem Junction the 600m Burslem Branch was closed in 1962 following a mining collapse but there are restoration plans. Festival Waters is a 16ha redevelopment site from Middleport to Etruria with offices, housing and leisure facilities. Burslem takes its name from the Old English man Burgheard and the British Celtic *lyme* or *lyne*, elm tree district. Cobridge was Bennet's Bleakridge, at the heart of *Clayhanger*. Shelton Bar steelworks (his Cauldon Bar Ironworks) were on both sides of the canal but closed in 1978, leaving the largest derelict site in the Midlands. The rolling mills had a roof over the canal, closed in 1995, and the last load was taken by *Shad* to the National Waterways Museum, Ellesmere Port. Etruria, named after what were thought to be Etruscan designs, was the site of the old Wedgwood works and it was

used for the 1986 National Garden Festival. The Festival Park has the UK's top water theme park, Waterworld, with pool, flumes, wave pool, rapids area, aqua assault course, cinema, bowling, ski slope and Quasar.

The Caldon Canal climbs away from Etruria Junction at this point. The Summit Lock at Etruria, an earlier name for Tuscany, had a toll house and a wooden roof over the lock until settlement reduced the headroom too much, door tops being at waist level. One of the locks has a fall of 3.99m, one of the deepest narrow locks around. Next to the second of the three Etruria Locks is the Etruria Industrial Museum, developed around the 1857 Jesse Shirley Bone & Flint Mill and including the last-surviving steam potter's mill, with the 1820s beam engine *Princess*. The former Etruscan gasworks were said to have the largest gas holder in Europe. Further across is the centre of Hanley (from the Old English *hean-lea*, high glade) with the Potteries Museum & Art Gallery containing the world's largest collection of ceramics and a Second World War Spitfire, designed by locally born aeronautical engineer Reginald Mitchell.

Stoke-on-Trent comes from the Old English *stoc*, dairy farm, and *Trent* is British Celtic for flooding river.

In 1487 the Battle of Stoke resulted in the defeat of impostor Lambert Simnel, who was then put to work as a palace turnspit after this last battle of the Wars of the Roses, bringing an end to the Lancastrians and, perhaps, medieval England. A hard-fought battle, it involved more combatants than Bosworth.

The Wedgwood Visitor Centre and museum has demonstrations, an art gallery and a restaurant with food served on fine Wedgwood china. One of the largest china factories in the world was established here after a move to this location from the city centre in 1940. Barlaston

A pair of bottle kilns in the centre of Stoke-on-Trent, restored in 2008 by a housing-estate developer.

Hall is a notable 18th century Palladian building constructed for attorney Thomas Mills.

Few towns are so clearly welcoming of canal traffic as Stone. There is obvious pride that this is where the canal began. A tree at the bottom of a canalside garden has a nestbox mounted with a television aerial, a juxtaposition that suggests they give their wildlife every possible amenity. Stone takes its name from the cairns on the graves of two 7th century Mercian princes, killed by their pagan father, King Wulfhere, for practising Christianity. The Crown Inn was designed in 1779 by Henry Holland for mail coaches and it was the coaching that led Defoe to feature the town in *Moll Flanders*. Just above the A520 crossing are a restaurant and the Star, the latter with its only entrance on the towpath at the bottom lock. The town has an annual food festival.

Where Amerton Brook enters, a track leads up to a lane to Hixon. On the other side of the river, Ingestre Hall was the Earl of Shrewsbury's Jacobean building, damaged by fire in 1882 and restored as an arts centre. The front was by Nash, the landscaping by Capability Brown and the adjacent St Mary's church of 1676 by Sir Christopher Wren. This is where Edward VII spent his holidays.

After Hoo Mill Lock, powerlines cross and there is a garden centre and tea rooms before Great Haywood Junction, where the Staffordshire & Worcestershire Canal leaves. The Trent & Mersey Canal now forms part of the Black Country Ring to Fradley.

Shugborough Park has been restored as a 19th century working estate. An 1805 farm is

Etruria Industrial Museum. In the spring the lock is surrounded by blossom.

Stone Lock, with its balance beam that is short in order to clear the road bridge.

the Staffordshire County Museum with farm machinery, working cornmill, demonstrations of farming methods and historic breeds of livestock. There are restored estate interiors, a kitchen, butlers' pantry, brewhouse, laundry, coach house, shops, domestic life exhibits, costumes, toys, crafts, steam locomotives and a cafe.

It was bought by William Anson from the Bishops of Lichfield in 1624 and since then has been the home of the Ansons, who evicted a village in order to expand their estate. William's great-grandson, George, became First Lord of the Admiralty in 1751 and repaired the neglect of the Royal Navy. His stay in Canton, during his four-year round-the-world voyage, resulted in the Chinese House. Other art treasures include 18th century French and English china,

French furniture, tapestry and paintings, silver, Vassalli rococo plasterwork and eight beautiful neoclassical monuments, including an 18th century monument by James Athenian Stuart. It is the earliest neo-Grecian building in the country. The hall of Shugborough, which means meeting place of witches and hobgoblins, is set in 3.6km² of parkland.

The iron bridge once carried a drive to Shugborough Hall with its pines, beeches and rhododendrons. The Essex Bridge was a late 16th century packhorse structure of sandstone. It is 1.5m wide externally and was the longest in the country. Fourteen of the 42 arches remain and it is now 94m long with a triangular cutwater at each pier. It was built as a hunting access bridge for the Earl of Essex to Cannock Chase, 50km² of heath and an Area of Outstanding Natural Beauty.

A Brindley aqueduct carries the canal on six brick segmental arches over the River Trent and this is the first time since Stoke that the river can be seen from the water, although the two have run side-by-side for 30km. Brindley's Bank takes its name from the canal's engineer. The Bloody Steps date from 1839 when Christina Collins was carried up to the Talbot Inn. She had been raped and murdered by the drunken Pickfords boat crew, with whom she had been a passenger; two were hung and one transported for 14 years for the crime. The sandstone steps, since replaced, were said to ooze blood on the anniversary and ghostly sounds and apparitions are still experienced. Another Rugeley miscreant was Dr William Palmer, publicly executed in Stafford in 1855 as a multiple poisoner.

After passing the Plum Pudding Inn and the Spode Cottage pub, the canal winds round the back of the church in Armitage with its Saxon font, its 1690 tower and the rest rebuilt in the 19th century in Saxon- and Norman-style. Inside is a loud 200-year-old organ originally from Lichfield cathedral.

Near Wood End Lock the canal reaches its most southerly point, turning sharply north-east to continue its journey. To the south can be seen the spires of Lichfield cathedral. To the north is the open farmland of the Trent valley. A stream runs northwards and it is the valley of this that causes the pronounced corner in the canal's route. A former airfield is seen through the trees to the east.

Shade House and Fradley Middle Locks drop the canal to Fradley Junction, locks here being the busiest in the country, accompanied by an 18th century Grade II workshop. The Coventry Canal joins opposite the Swan Inn, built to serve the boat trade.

The Trent & Mersey Canal becomes part of the East Midlands Circuit, the flight continuing with Fradley Junction, Keeper's, Hunt's and Common Locks. Several locks have corners that are not easily accessible, which have been turned into small gardens of flowers or shrubs by local residents. The crane is an import from Horninglow Basin.

The Trent is braided here and the canal passes through another lock and then uses river channels for over a kilometre. Alrewas takes its name from 'alder wash', a swamp of alder trees used for basket making.

From Wychnor Lock the canal joins a major traffic corridor, opened up by the Romans with

Spode House, now Hawesyard Hall, has given its name to fine porcelain.

their Ryknild Street. This has now become the A38, a dual carriageway forming the major link between the West and East Midlands. The canal follows the A38 for 15km to Egginton, often right alongside but otherwise just far enough away for the roar of traffic to die back.

Across the fields, a 16th century battlemented church tower locates Barton-under-Needwood at the foot of Needwood Forest. This was a royal forest with its only access from Barton Turn. The Barton Turns Inn and Barton Marina with its shops accompany Barton Lock, followed by a junction of the B5016 with the A38, one slip road looping over the canal.

The Bridge Inn at Branston would seem the ideal place to stop for a ploughman's and pickle but the menu seems to major on variations of curried chicken while Branston pickle is now made elsewhere. On the other side of the canal there are extensive views to the ridge of Tatenhill Common, where a motorcycle trials route winds its way up and down the hillside.

Burton upon Trent is dominated by a large brewery tower. Perhaps it also had a tower when it was the Old English *burh-tun*, a fortified village.

For the only time before its last lock, the Trent & Mersey Canal makes brief use of the River Trent itself.

This is Britain's brewing capital, often marked by the odours of brewing, an activity thought to have been started by monks in the 13th century, using water thar was very clear because of its high gypsum content, still supplied from wells and springs. The first brewery was established in 1708 and there were a score at one time. Burton's most popular product is India Pale Ale, initially made for export only, until a boatload sank and was released on to the home market by the underwriters.

At Horninglow Basin, where there was a salt warehouse until the 1970s, a large canal mural decorates the side of the embankment of the A38, now back alongside.

The canal moves from Staffordshire to what is soon to become much more rural Derbyshire as it crosses the first of the two channels of the River Dove. The canal uses a dozen low brick arches, at 73m the longest aqueduct on the canal, built by Brindley, while the A38 uses a classic-style bridge, despite the roads being of almost motorway-standard.

The canal parts company with the A38 here and the route becomes more peaceful. There is a small works at the approach to Willington, where the A5132 and B5008 cross, the latter with a picnic site. The 585-berth Mercia Marina,

the country's largest marina, has been built in a former trout angling lake.

After Stenson Marina, the Bubble Inn (perhaps named after the lock discharge boil) and Stenson Lock (which has a 3.8m fall and is the first wide-beam lock at this end of the canal) there are occasional extensive views southwards across the Trent Valley.

The 18th century stone-and-brick Arleston House farmhouse is followed by the Ragley Boat Stop public house. A minor road crossing the canal to Barrow upon Trent has a very odd parapet line, although it probably looks fine to road users.

The following bridge crosses from Swarkestone Lows to Swarkestone and the Lowes Bridge, a five-arch structure that crosses the Trent, continuing across the meadows for a further 1.2km with another dozen arches, making it the longest stone bridge in the UK. From the 18th century, part remains from the 13th century, having been built by two sisters of the Harpur family whose fiancés were drowned trying to ford back through the flooded Trent after being called across to a meeting of barons. There was a Civil War battle to control it and it is where Prince Charlie's Jacobite army of 1745 abandoned their attempt on London and began their retreat to Scotland.

The freight railway crosses at Chellaston and long views open up to the south. Willington powerlines mark where a further former railway bridge crosses, now carrying the Derby Cycle Route, just a little less rural in the present day than in the times when *diuraby* was from the Old Scandinavian for deer, giving the future city its name.

Red-brick waterways cottages flank the grey-stone Georgian Shardlow Hall. Shardlow, with the B5010 crossing the canal in the centre, is a canal jewel, one of the first inland ports, dating from 1797, and formerly the largest. After Shardlow Lock is the Clock Warehouse, a four-storey corn mill of 1780 with a barge entry point, now converted to a popular family public house. Shardlow Heritage Centre is in the Salt Warehouse and there are maltings and the Malt Shovel Inn of 1799. Everywhere there are warehouses plus a feedstuffs mill, a maltings and a brewery. In 1839, 140 boatmen at the port called for the canal to be closed on Sundays for religious reasons. The canal company took up the call and tried unsuccessfully to get Parliament to apply this to all canals and railways. The village claims the largest number of public houses in the country for its population size.

Moored among the many wharves are large numbers of narrowboats, likely to leave a film of spilt fuel on the water surface. The Lady in Grey restaurant is approached through a hole in a wall but the New Inn at Great Wilne is less coy.

The toll wharf at Swarkestone.

30 Macclesfield Canal

The Macclesfield Canal was one of the later canals to be built and one of the last narrow canals, being opened in 1831. It was designed to act as a shortcut around part of the Trent & Mersey and Bridgewater Canals. Thomas Telford selected the line and used bold embankments and cuttings to produce a shorter journey time, rather than using the more economical construction method of following contours. This produced a canal with dramatic views, helped by the proximity of the Peak District, whereas the Trent & Mersey had kept to more even and lower ground further to the west. Notwithstanding the extensive cut and fill, engineer William Crosley managed to cut a couple of kilometres off Telford's line. He produced landscaping that was advanced for its time and often the surrounding land blends up or down to the canal edge in a way not achieved by most of our motorways.

To enthusiasts of canal architecture, the most attractive spot on the canal comes immediately at Marple, with stone buildings adjacent to a stop lock, which was probably never used because of good relationships between the canal ↗

The stop lock at Marple.

companies, particularly over water supply. The whole ensemble is framed at each end by a pair of immaculate roving bridges, of which the canal has several excellent examples.

The canal quickly breaks into open country although it is just possible to glimpse the spread of spires, gas holders and tower blocks of the south of Manchester through gaps in the hedge on the right. The noise of jets using Manchester Airport can be heard.

Just before leaving High Lane, the Bull's Head is passed on the right while the Dog & Partridge offers food 200m up the road the other way.

An aqueduct carries the canal over the railway on its first high embankment and gives views through to Disley, a castle-like structure in Lyme Country Park and the Peak District National Park, which runs parallel to the canal for much of its length and comes within 600m of the canal at one point.

Arrival at Bollington is marked by the Clarence cotton spinning mill, which has been transformed into accommodation, shops, Canalside Community Radio and a Discovery Centre. This time it is of stone but still has the hoist over the canal. An aqueduct takes the canal over the river. Strangely, Bollington lies on the River Dean while Macclesfield is on the River Bollin. The Grade II Adelphi Mill for cotton and then silk, including fabric for parachutes, houses commercial premises. The aqueduct over the B5091 has the stone of its semi-circular arch curved into its wingwalls.

High-level passage gives views across the rooftops to the right while peaks such as Nab Head, Kerridge Hill with the 5.5m Grade II White Nancy Waterloo monument and Tower Hill dominate to the east. Bollington was once supported by its cotton mills and another stands silent on the right of the canal, the four turrets festooned with aerials on top of its tower mocking its quietness. A local resident was Terry Waite.

Distance
42km from Marple Junction to Hall Green
Highlights
Marple, one of the most attractive spots in terms of canal architecture
The Goyt and other cotton and silk mills
Bosley Locks and the Cloud
Exemplary Elizabethan Little Morton Hall
Navigation Authority
Canal & River Trust
Canal Society
Macclesfield Canal Society
www.macclesfieldcanal.org.uk
OS 1:50,000 Sheets
109 Manchester
118 Stoke-on-Trent & Macclesfield

Hovis Mill, the original source of flour for Hovis bread and later for the wrappers, stands beside the canal as high-class apartments, its moorings now used by Freedom Boats. There were 71 mills in Macclesfield for the silk industry, including Paradise Mill which closed in 1981. This was the world's leading production centre of silk and women working in the industry were nine times as likely to die of lung diseases in the mid 19th century as people in agricultural districts. Macclesfield

The Cloud rises above the Dane Aqueduct.

Heritage Centre features silk and Sunday schools. A heavy retaining wall flanks the hill on the left, repaired after a 2008 collapse, its base surrounded by heather. It was repaired again in 2012, local residents having had their cars trapped in their garages for months because of the settlement. Bridge 41 is much higher and wider than other bridges, without a constriction of the canal, but it is built in similar style and probably took a tramway across.

At Oakgrove, the Fool's Nook provides food and beer. The low lifting Royal Oak Bridge crosses the canal. The hillside on the left is wooded for a while. A large house by the canal has its opening windows plain while the others are leaded, an effect that looks incongruous.

The canal has been lined with reedmace sporadically but as the ground falls away to the right, taking the canal with it, the rushes close

in and narrow the otherwise wide canal for the first time. The canal was designed so that all the locks come together in a flight of 12 at Bosley, built with stone from the Cloud. They have mitred top gates because of the depth of the locks. Side pounds exist at all except the top lock, which now has a garden instead. Both these and the locks themselves, all listed buildings, are built on an embankment in places. With the rapid fall of the ground and the adjacent disused railway embankment, there is some dramatic topography, including Sutton Common with its mast. Overlooking it all is the Cloud, a 343m hill with the remains of ancient earthworks beyond the bottom of the flight. Part of the craggy appearance is from a quarry that provided stone for the canal and then the railway.

Milestones are met before and after the railway passes over the canal. An extreme example of Telford's no-nonsense line comes

Passing under Scolar Green's roving bridge.

in the form of a 2km dead-straight length to the east of Congleton. Unfortunately, the same thinking was not employed at bridge 75, where the concrete edging of a flyover does not line up with the stonework of the original of the three bridges sharing this number 75, and which forms a chicane for narrowboats. The third bridge carries the railway.

At the other side of HJ Lea Oakes' cattle meal mill, perhaps Britains longest-established manufacturer, the Railway public house offers the landlady's home-made pies. The Queen's Head is keen to promote its food, too, and is approached from bridge 76 which, like the bridge beyond, is a fine roving bridge.

Mow Cop dominates the eastern skyline. Its summit is capped by Randle Wilbraham's folly in the form of a ruined castle dating from 1754, inside which a 14-hour prayer meeting for 4,000 was held in 1807 by Hugh Bourne, resulting in the birth of Primitive Methodism. The Mow Cop Killer Mile is run annually to the summit from the canal.

The 18th century brick Ramsdell Hall, standing back from the left bank of the canal, overlooks a magnificent spread of lawn – a meadow-sized putting green. Looking back on the other side of the canal, if the trees permit, it may be possible to see Little Moreton Hall just over a kilometre away. Built between 1559 and 1580, Little Moreton Hall is surrounded by a moat and is probably the best example of Elizabethan half-timbering in existence. It has a fine collection of pewter and oak furniture, a 21m Long Gallery with original panelling, a 16th century chapel, an Elizabethan herb garden and a 1688 knot garden.

Kidsgrove appears on the hill ahead. At Red Bull, an aqueduct carries the Hall Green Branch over the Trent & Mersey, one of only three such canal flyovers in Britain. A sharp left turn by Red Bull Wharf brings the Hall Green Branch parallel to the Trent & Mersey, which it joins at Hardings Wood Junction after another 600m, during which time the Trent & Mersey has climbed through two more locks to meet it. Many boats have cruised back and forth along this stretch of the Trent & Mersey, trying to find the entrance to the Macclesfield Canal on the north side.

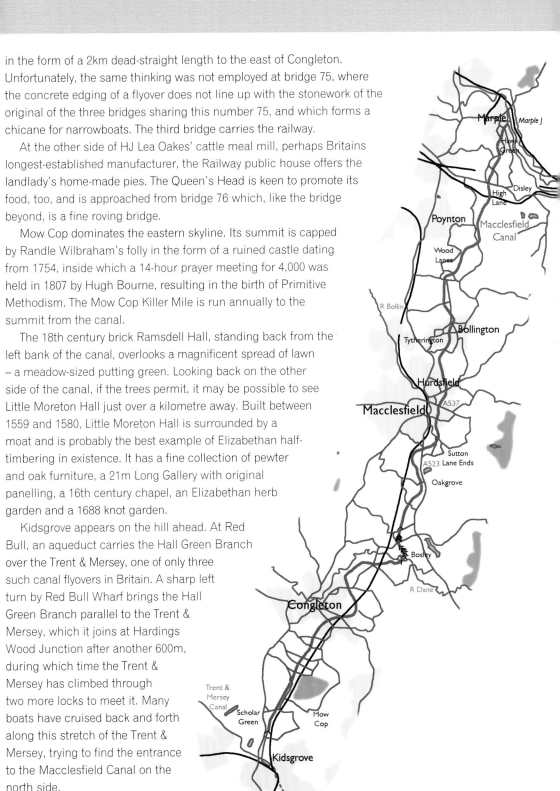

31 Caldon Canal

Although a branch of the Trent & Mersey Canal, the main business of the Caldon Canal was self-contained when it was constructed in 1778. Its purpose was to bring limestone from the quarries at Caldon Low to the industrial area at Etruria, site of Josiah Wedgwood and Thomas Bentley's factory producing Jasper Ware, red-and-black figure porcelain vases and cameos. The siting is where the Etruria marl meets the coal measures of the North Staffordshire coalfield, the marl and coal being used 1:6 in the production of the porcelain. It was the canal that established the commercial importance of Stoke-on-Trent. It carried pottery traffic until 1995, the last regular commercial traffic on any narrow canal.

The area has since been turned into the 3ha Etruria Industrial Museum with a waterside public house, museum, steam beam engine from about 1850 and the Etruscan Bone & Flint Mill with its iron-banded brick chimney, erected in 1857. Also included are the canal maintenance buildings and dry dock at the junction with the Trent & Mersey, just above Etruria Top Lock.

The Caldon Canal has one of the most dramatic changes of scenery over its length of any of the British canals. The start is past new housing, fronted by a 1990 statue of Brindley, who

Above Fens Lock at Stockton Brook.

died after getting soaked while surveying the canal. An arched bridge has X-braced parapets, perhaps inspired by Horseley Ironworks bridges.

The first two of the 17 narrow locks come as a staircase at Bedford Street. Like the following locks, they have large side weirs and have split footbridges across their lower ends, supported by decorative ironwork.

The line cuts through the ornamental Hanley Park with its ornately decorated canal bridges, bandstand and other public facilities before emerging to give the first views of the moorlands, sweeping round to the transmission mast at Fenton.

At Sneyd Green (Bennett's Toft End) the canal turns sharply right just by the Foxley public house and the blocked-off end of the 800m Foxley arm, disused since 1934, having served a colliery and ironworks.

This long pound ends at Norton-in-the-Moors, with lock 4, Engine Lock, taking its name from a steam-powered beam engine that pumped water from a nearby mine. The conurbations of Stoke-on-Trent fall back, too, and open country is met for the first time.

The canal now begins to blend into the landscape in a most agreeable fashion as fields slope gently down to water level rather than being cut into by the canal. This is particularly so at Stockton Brook, where irregular hilly fields dominated by clumps of deciduous trees tumble down to meet the canal as it climbs its final flight of five locks, accompanied by sculptures showing features of local industry and life. These five begin with Waterworks Lock, which takes its name from the Victorian waterworks alongside, and follow with Fens Lock, Railway Lock where the line crosses over, Road Lock by The Sportsman pub and Hollybush Lock, which has its large side weir discharging beneath a set of old canal buildings. The canal is now 148m above sea level and takes off with a broad straight pound instead of the narrow twisting route taken so far.

Distance
41km from Etruria to Combridge
Highlights
Etruria Industrial Museum
Cheddleton flint mills
Navigation Authority
Canal & River Trust
Canal Society
Caldon & Uttoxeter Canals Trust
www.cuct.org.uk
OS 1:50,000 Sheets
118 Stoke-on-Trent & Macclesfield
119 Buxton & Matlock
128 Derby & Burton upon Trent

This is interrupted at Endon by a circular island in the canal, the base of a swing bridge for a former light railway. The main line used to continue through what is now Endon Basin but was diverted to the south in 1841 to accommodate the railway and feeder from Rudyard Reservoir.

Endon, which is dominated by a creeper-covered house on a hilltop, obtains minor fame for two days in late May each year as being the only village outside Derbyshire where well dressing is practised.

The occasional rock outcrops, brick bridges being progressively replaced by stone and cows feeding in woods at the base of a steep bank, lead

The Bedford Street staircase at Etruria.

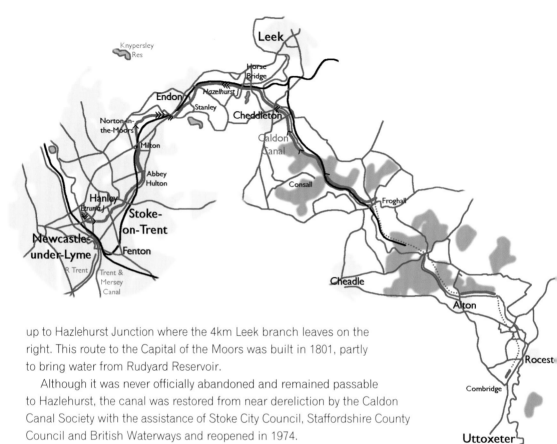

up to Hazlehurst Junction where the 4km Leek branch leaves on the right. This route to the Capital of the Moors was built in 1801, partly to bring water from Rudyard Reservoir.

Although it was never officially abandoned and remained passable to Hazlehurst, the canal was restored from near dereliction by the Caldon Canal Society with the assistance of Stoke City Council, Staffordshire County Council and British Waterways and reopened in 1974.

Beyond the Hollybush at Denford, the canal swings southwards into the valley of the River Churnet and some of the finest scenery in the Midlands, deep wooded valleys with overhanging crags.

At Cheddleton comes the first of several unusual side weirs with the towpath taken across the face of the weir while a narrow footbridge adjacent enables the man leading the tow horse to keep his own feet dry.

Cheddleton is renowned for its mill complex with buildings dating back as far as the 13th century. Its pride is a magnificent pair of undershot mill wheels. One of these flint mills was probably designed by James Brindley. Restored in 1967, the mills were used to grind flint from the English Channel before sending it up the canal to the Potteries. Other exhibits include a Robens steam engine, a haystack boiler, a model Newcomen engine, a section of plateway and lime and ochre kilns. Overlooking the complex is a church with pre-Raphaelite art and stained glass.

Beyond a wooden building across the canal, two locks drop past Castros restaurant and a farm-gate factory before the canal reaches the station for the restored Churnet Valley Railway, which follows the canal, and the Staffordshire Way footpath comes alongside.

Normally dark and slow, the river winds down to Consallforge, once an extensive water-powered ironworks with restored limekilns, where the canal breaks away from the river past the Black Lion and into woodland. There is barely room for the railway to squeeze in between the canal and river here and a wooden footway takes the towpath under the railway bridge.

Cherry Eye Bridge, with its pointed, stone arch, marks the start of Rueglow Wood, with its picnic area and mooring and a horse escape ramp. This reach, with idyllic scenery, has been relined in concrete. Thomas Bolton's copperworks have been demolished.

At Froghall, the canal turns sharply into the low and irregular 69m tunnel with its roof sagging in the centre. Froghall Wharf, the terminal basin, has a canal craft centre, horse-drawn narrowboat trips and a picnic area. It has ruined limekilns and, when constructed, was connected by horse-drawn tramway to the quarries at Caldon Low.

In 1811, the canal was extended southwards to Uttoxeter. The Uttoxeter Canal, as the new part of the Caldon Canal was sometimes called, was built as a blocking move to prevent a broad-beam canal being built to rival the Trent & Mersey. It ran for a further 21km with 17 locks. In 1846 the North Staffordshire Railway bought the Trent & Mersey Canal, of which this was a branch, and converted most of the canal extension to a railway. This has, in turn, been closed although Staffordshire's only Italianate station has been restored at Alton and there are restoration plans for the Churnet Valley Railway. The first lock on the Uttoxeter Canal at Froghall has been restored, giving access to a lower basin.

For 2km at Alton the bed has water, albeit shallow. Although it skirts the southern boundary of Alton Towers, Britain's largest amusement park with the highest number of paying visitors – over 27,000,000 per year – the 1980 venue remains hidden and nothing more than the screams of punters is heard. On the other hand, the castle, with Pugin's mixture of Scottish and French Gothic restyling, is the dominant building, high on its crag on the south side of the valley.

Trip boat moored at Froghall Wharf.

32 Bridgewater Canal

The Bridgewater Canal was one of the engineering wonders of the age, the first Industrial Revolution canal to be built away from the course of an existing river, beginning the canal-building era. To this day it remains the largest British water project to have been financed by one person. The Duke of Bridgewater turned his attention to business after being jilted and employed John Gilbert and James Brindley to build the canal. The canal was begun in 1759 and reached Castlefield by 1765. A wide canal, it followed the 27m contour and had no locks, despite some long straights, except at Runcorn, to which it was later extended to join the Mersey estuary, opening throughout in 1776.

It was bought by the Manchester Ship Canal Company in 1885, the builders of the biggest modern canal, which superseded it, although it continued to carry commercial trade until the early 1970s.

Today it forms part of the Cheshire Ring from Castlefield Junction to Preston Brook.

The canal is fed by the River Medlock, emerging from under a low bridge. It starts in Castlefield Urban Heritage Park, Britain's first such, with the Museum of Science & Industry including a Beyer-Garratt engine, a Shackleton bomber and a power station condenser. Manchester Central, north-west England's largest conference and exhibition centre, is located in the former Central Railway station. The Manchester Hilton includes apartments above the 25th floor. Manchester takes its name from the British Celtic *Mamucium* after the breast-shaped hill on which it was built, that hill now being lost among a forest of buildings.

The highly distinctive state-of-the-art Merchants' Footbridge has crossed the Bridgewater Canal since 1995. Curved in both plan and elevation, it is supported by a single arch, like a bowstring girder but leaning upwards and towards the inside of the curve at 60°. Its structural integrity depends on torsion, a concept more commonly used in aircraft design than in bridges. It is 67m long. The more traditional kind of bridging is seen at its best in the Castlefield Viaducts, massive Victorian girder structures that carry everything from the line to Liverpool to the Manchester Metro, which follows the canal to Timperley. To build the viaducts involved evicting hundreds of slum dwellers from Alport Town without compensation.

Pomona Docks, on the Manchester Ship Canal, were for coastal and perishable traffic. They were opened in 1894, the third-largest in the country, amazing for a location so far inland. No 3 dock was a Ro-Ro terminal, with Colgate's Wharf opposite. The land between the two canals was previously the pleasure gardens. Pomona Palace was the location for political meetings with Disraeli speaking to 28,000 people in one hall, not to mention many concerts and shows.

Castle Quay with its drawbridge.

Distance
45km from Castlefield to Runcorn
Highlights
Castle Quay and Castlefield Viaducts
The Lowry and Imperial War Museum at Salford Quays
Old Trafford, the home of English football
Navigation Authority
Manchester Ship Canal Company
Canal Societies
Bridgewater Canal Trust
www.bridgewatercanal.co.uk
Runcorn Locks Restoration Society
www.runcornlocksrestoration.org
OS 1:50,000 Sheets
108 Liverpool
109 Manchester

The botanical gardens were opened in 1831, John Dalton selecting it as the cleanest suburb in Manchester after a tour in which he wiped leaves with his handkerchief to check for pollution. In 1857 it was the site of an Art Treasures exhibition with an even larger Crystal Palace exhibition in 1887. It was all cleared to build the docks, a process repeated when the docks became an Enterprise Zone for redevelopment in 1981. Now they are better known as Salford Quays with such amenities as The Lowry theatre and gallery and the Imperial War Museum North. It is also the home to MediaCityUK, the principal tenant of which is the BBC.

One of the more dramatic buildings has a chain and hook on each side, each link of the chain being several metres high, the chains being freestanding and leaning in towards the building. The adjacent arched bridge crosses the Manchester Ship Canal. Throstle Nest bridge must be the most clearly named bridge on the canal system. Trafford Park was the workshop of the world in the world's first planned industrial estates. It was built on a former deer park owned by the de Trafford family since the time of Cnut and sold in 1896. Activity in the complex peaked during the Second World War, when 75,000 people were employed there, producing Lancaster bombers, Rolls Royce aero engines, the PLUTO pipeline, penicillin, DDT and dyestuffs to be used for camouflage.

To football fans, Trafford Park means the home of Manchester United Football Club. It is one of the spiritual homes of the game, the 1910 Theatre of Dreams, Old Trafford Stadium, the UK's second largest, reaching almost to the bank of the canal. The Old Trafford cricket ground lies a little to the south-east.

Stretford Junction or Waters Meeting brings in the Stretford & Leigh Branch, the first section of the canal to be built, and the route follows that original line briefly, passing playing fields

The Beetham Tower, Pete Waterman's chapel-turned-recording studio and a recent footbridge.

after being crossed by another freight line and the Manchester Piccadilly to Liverpool railway. The 1761 terminus lay between Stretford gasworks and the A56.

Stretford Wharf, by the cemetery, had steps up out of the canal so that any horse that fell in could be recovered. There were stables close by at the packet changeover point, connected by the lengthsman's cottage. His most important task was to watch Barfoot Bridge over the River Mersey, hence Watch House, which gives its name to the boat cruising club that occupies the building today. The water is generally opaque. The canal company built a weir upstream to pass floodwater away from Barfoot Bridge, an aqueduct with a single 20m span over the Mersey.

The valley is followed by the Trans Pennine Trail and contains Sale Water Park and Chorlton Water Park, with activities in former gravel borrow pits for the M60, which now crosses with Junction 7 just west of the canal.

Manchester University Boat Club is on the corner by the Bridge Inn. Canoeists travel at speed, looking the opposite way, and their blades sweep almost the whole width of the canal.

The 19th century St Paul's church stands prominently by the canal – perhaps too conspicuously as it was hit by an incendiary bomb in 1940, destroying the roof and leaving scars that still show on some of the pews. The cemetery that follows was once fashionable and its occupants include Richard Pankhurst and James Joule.

The Manchester to Chester railway leaves the line of the canal after Timperley station. Once the railway was opened in 1849, the packet boats stopped here and discharged their passengers so that

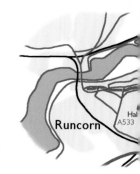

they could travel on to Manchester by the much quicker trains. The line was electrified in 1930. A freight line and oblique bridge that previously carried another line both cross.

The Roman road from Manchester to Chester didn't bother with the A56's detour to Altrincham. It crossed the canal's line somewhere near the 1897 factory of Linotype, with its elaborate clocktower, a factory that has played a vital role in the development of printing.

The 1km² Dunham Park has 150 fallow deer. It has 180 species of beetle, old trees not being felled nor fallen timber removed. An obelisk, dating from 1714, terminates one of the tree-lined avenues in the park. Dunham Massey Hall is a Georgian rebuilding of a Tudor house, the family home of the Booths and then Earls of Stamford. It has a superb collection of 18th century furniture, Huguenot silver and portraits of the Booth and Grey families, including Lady Jane Grey, Britain's briefest-reigning monarch, on the throne for just nine days. Thirty rooms are open to the public, including the great hall, dining room laid out for a 1913 Edwardian shooting party, vast kitchen, library, laundry, 18th century stables (containing a restaurant) and Elizabethan cornmill, converted to a sawmill in the 19th century and working most Wednesdays. Around the park are an orangery, well house, slaughter house, Edwardian water gardens, one of Britain's biggest winter gardens and 160 varieties of roses.

The Bridgewater Canal crosses the River Bollin on its highest embankment, 10m above the river, not so much an aqueduct as a concrete tower block with a flat arch cut into the bottom and a rapid passing below. In 1971 a breach occurred and the water level had dropped 360mm in Manchester before the stop planks could be placed. Meanwhile, the canal water had carved a 30m gorge in the embankment, carrying sand, gravel and rocks into the river.

A new concrete and steel trough for the canal was opened two years after the event and the Bridgewater Canal Trust was set up to finance proper maintenance of the canal.

At the river, the canal moves from Greater Manchester into Cheshire and passes Little Bollington, the diminution being added after local government reorganisation in 1974 to avoid confusion with the Bollington on the River Dean. Bollington coal wharf and warehouse were sited here.

Whitbarrow Aqueduct had an overflow to the Slitten Brook. A Victorian park is sited in the valley, which had a slitting mill that flattened iron bars into strips for making into nails, tools and barrel hoops. The mill was later used for woollen manufacture after 1800. There was also a corn mill. Assorted watering holes include the Golden Fleece, Saddlers Wine Bar and Lymm Conservative Club.

The A56 crosses again at Ditchfield Wharf, where there is another crane with stop planks and an iron bar to protect the bridge from towlines. More noisily, the M6 also crosses as it rises on to Thelwall Viaduct to cross the Manchester Ship Canal. The viaduct has been dualled, the first viaduct having suffered badly from salt corrosion. Around 15 tonnes of salt is applied to the structure per year to keep it free of ice. The replacement cost of the viaduct was £10,000,000 or £27,000 per tonne of salt applied.

Thelwall Grange, built in quieter Victorian times, is now a nursing home. Further along the hillside is Massey Hall, now a school but previously the house of the owner of Rylands Wire Works. Thelwall was founded in 923 by Edward the Elder, with a camp to guard the Mersey from the invading Danes.

The top of a green marine navigation buoy at Cliff Lane Bridge Wharf draws attention to S&A Marine in a former tannery owned by the Co-operative Wholesale Society. Soon after, the A56 crosses on a heavy concrete bowstring bridge.

A notable building in Grappenhall is the 16th century St Wilfred's church by the canal, some parts of which date back to the 12th century. In 1874 it was enlarged and restored. At the gate there are stocks. Inside there is a Norman red sandstone oblong font, a 13th century dugout chest made from a tree trunk and a 1275 effigy of Sir William Boydell. Some of the stained glass is very old and at one time it was said to have more stained glass than any other church in Cheshire. There is a 180-year-old sundial and a stone cat above the west window. Lewis Carroll's father preached here at times and would have been accompanied by his family, giving credence to the suggestion that the cat was to become the Cheshire Cat in *Alice's Adventures in Wonderland*.

Stockton Heath warehouse and wharf have semi-circular steps that served as the interchange between fast packet boats from Manchester and Runcorn and stage coaches for Warrington or the south. The London Bridge takes its name from the A49 London Road which crosses, the Roman road from Middlewich. It was home to the *Duchess-Countess* packet boat, which took five hours to reach Manchester and carried a blade on the front to cut any towlines of other boats that were not dropped in time. A model of the boat is to be found in the National Waterways Museum, Ellesmere Port. Stables for the packet boat horses were in the house of the bank rider, now occupied by Thorn Marine. The dovecote is a recent addition.

The Delamere Way footpath follows the canal from its Stockton Heath end towards Frodsham, passing the Fox Covert cemetery and older Baptist one beyond (where Oliver Cromwell once worshipped) and leaving across the golf course. On the other side of the canal is a cricket pitch. Hill Cliffe Reservoir contains 45,000m^3 of water for Warrington. Hillfoot Farm's 17th century barn has an arched end and round holes for pitching hay or straw down from the loft. Water from Appleton Reservoir is supplied to Warrington at a rate of 2,200m³/day through a pipe carried on a wooden frame at Hough's Bridge, a crossing named after a local family.

Another local family of great significance at Higher Walton were the Greenalls, noted for their brewing. Two generations of the family rebuilt the village, the Victorian and Edwardian houses now mostly having been sold to tenants. The Gothic revival church of 1885 was built by Paley and Austin for Sir Gilbert Greenall and he had Walton Hall built in 1836, these days featuring a children's playground, pitch and putt course, outdoor chess, museum, children's zoo and gardens laid out by the first Lady Daresbury. The canal runs in a broad sandstone cutting.

Moorfield bridge has another stop plank crane and winding hole, on a section of canal above two railway lines and with views across the Mersey to Fiddler's Ferry power station. Up the hill, beyond the potato fields, is the tower of All Saints church at Daresbury, the church where Lewis Carroll's father was the vicar, something clearly indicated by his characters in the stained-glass windows. Many boat crews were not prepared to come up

to the church, however, so the vicar converted a mission boat at Preston Brook for their use.

A more conspicuous tower, these days, is the white one of Daresbury Nuclear Physics Laboratory, housing the world's largest tandem Van de Graaff generator. The laboratory also had the first synchron radiation source for research, cooled by water drawn from the canal.

The short Preston Brook Branch connects through Preston Brook tunnel with the Trent & Mersey Canal. The plan was that transhipment should be done at Middlewich but, by an unfortunate error, the tunnel was just too narrow for Mersey flats to get through so it all had to be done here, to the considerable financial gain of the Duke of Bridgewater. Two kilometres of wharves, warehouses, offices, houses and stables were set up from the tunnel on to the main line towards Runcorn, forming one of the busiest inland ports in the country, later becoming a canal/railway interchange with the Warrington to Crewe line passing under the end of the line to Runcorn.

The broad sandstone cutting past Walton Hall.

One of the few remaining canal buildings is Stitts warehouse, occupied by Pyranha, one of the world's leading canoe manufacturers and source of much original thinking in canoe design. Pyranha were also behind such landmark events as the descent of Everest and the majority of Richard Fox's world slalom championship successes. At one time, Norton warehouse covered the canal arm here but now it is open.

A marina for 300 boats was built opposite in 1974 and the towpath diverted across the canal and round the back of what is now Pyranha's site. From here the canal forms the northern boundary of Runcorn's housing.

A canalside notice invites water users to visit the remains of Norton Priory, an Augustinian centre from 1134 until the Dissolution in 1536, to which Henry VIII had given 30 oaks in 1516 to repair fire damage. The museum has the most extensive display on medieval monastic life in Britain and features the largest excavation of its kind in the country. There is a church, 12th century undercroft, chapter house, dormitory and cloisters display, 6ha of woodland and the gardens of a 16th century house, summerhouses, Victorian rock garden and stream glade with azaleas, Georgian walled garden, museum, shop and refreshments. It took a five-year legal battle to buy land from the owner of the priory to complete the canal.

Runcorn has a network of dedicated roads for single-deck buses and one crosses the canal from the direction of Castlefield, where the remains of the castle are to be found on top of the hill. The disused Barge, beside the canal, was attacked by an arsonist in 2011.

An industrial estate lies between the canal and the Manchester Ship Canal, hidden by the embankment of the A558 Daresbury Expressway and the A533 Bridgewater Expressway, which crosses and then runs alongside. Watering holes come quickly: the Royal Naval Association, the Grapes Inn, the Navigation and the Egerton

Arms. There were also various commercial premises: Astmoor wharf and tannery, Highfield tannery, Halton Road gasworks and wharf and the Bridgewater foundry.

The Sprinch was a loop containing Runcorn dockyard, left after a bend was straightened in 1890. When a minor road was dualled to form the new A533 a further section of the loop was lost, leaving two spurs. Victoria Dockyard was the main one on the canal, building and repairing the Bridgewater Company's extensive fleet.

Stitts warehouse, the last remaining of the canal transhipment facilities at Preston Brook. It is now home of Pyranha, Britain's largest canoe manufacturers. The water tower is visible on the hilltop.

Bridgewater Motor Boat Club, formed in 1951, the oldest on the canal, use the crane every Sunday to remove the gate from the large dry dock.

There were two flights of locks down to the Manchester Ship Canal. The New Line Flight of ten double locks was closed in the 1960s and has been built over, the lock gates being reused on the River Avon. The Old Line of five double locks has been preserved. When the second Mersey crossing is built, it is hoped that the flight will be restored, creating a second Cheshire Ring via the River Weaver.

Leeds & Liverpool Canal: Leigh Branch and Bridgewater Canal: Stretford & Leigh Branch

The Leeds & Liverpool Canal was the last of the three trans-Pennine canals. The Leigh Branch was opened in 1820. It runs south-east across Greater Manchester from the main line at Ince-in-Makerfield to join the Stretford & Leigh Branch of the Bridgewater Canal, thus avoiding craft having to cross the Mersey estuary to travel south.

The junction with the main line at Ince-in-Makerfield is just above Wigan Pier in the valley of the River Douglas. Wigan power station was demolished in 1989 but had been supplied along the branch until 1972 by short boats. These craft were characterised by ornate baroque scrolls, flowers and birds on the bows and stems and by square wooden chimneys, which were prone to catching fire.

The two Poolstock Locks are the only locks on the canal and their size determined the 18.9m x 4.3m of the distinct variety of wide-beam barges

Plank Lane lifting bridge.

used on the main line and on the Bridgewater Canal. Its wooden construction was based on Mersey and Weaver river barges with crew accommodation in fore and aft cabins below deck. Its maximum load of 45t was large and this contributed to the prosperity of the canal, even resisting railway competition at first. Steam engines were used from 1890, especially the V-shaped twin units of the Leeds & Liverpool Canal Carrying Company in Wigan, and they worked in pairs as motor and butty, the last one being launched in 1936 and the design finally being withdrawn in the 1950s.

One Poolstock lock was built with a wooden chamber to deal with subsidence problems. Both have penstocks that open sideways instead of vertically as would usually be the case and the lower lock has windlasses with chains to pull the balance beams shut instead of relying on leg power, which would normally be enough. It also has a footbridge with baffle plates to resist wheeled traffic.

Abram begins suddenly with houses right beside the canal but then draws back to leave the hillside on the left as grazing for horses, overlooked by a blackened but substantial stone church. To the right there are long views across the valley of the Glaze Brook. The canal runs on embankment, at times, with low ground on both sides. The Dover Lock Inn draws attention to a series of bank undulations that were once Dover Top Lock and Dover Low Lock, these two being eliminated when subsidence made them unnecessary.

Other redundant structures include Edge Green Basin, seen as a blocked-off inlet on the left, and a railway line that once crossed close by. Plank Lane Bridge was a swing bridge carrying a busy road and opened mechanically by a bridge keeper, until it was replaced by a troublesome, electrically operated lifting bridge.

The canal takes a rather odd line on its approach to Leigh, a series of wiggles that are

Distance

12km from the Leeds & Liverpool Canal main line to Leigh

17km from Leigh to the Bridgewater Canal main line

Highlights

Astley Green Colliery Museum

Worsley Basin Pocket House and entrances to the Duke of Bridgewater's mines

Barton Swing Aqueduct

Navigation Authority

Canal & River Trust

Manchester Ship Canal Company

Canal Society

Bridgewater Canal Trust

www.bridgewatercanal.co.uk

OS 1:50,000 Sheets

108 Liverpool

109 Manchester

so slight it is not clear why a straight line was not chosen. Beside the canal is a brick warehouse that has been restored as the Waterside Inn. The Moonraker is close to the canal.

The Leigh Branch meets its Bridgewater Canal equivalent on the far side of the A572 bridge in Leigh. Leigh comes from the Old English word *leah*; glade.

The Bridgewater Canal was the first of the modern canals to be constructed entirely independently of rivers. The first section was opened in 1761 and it began the era of canal building. A contour canal with no locks, it was a broad canal with many features that would not have seemed out of place later and which served as a model for many canals that were to follow.

It was constructed to bring coal from the mines of the 3rd Duke of Bridgewater at Worsley to the centre of Manchester. Construction of the new canal immediately halved the price of coal in Manchester, made a fortune for the duke and made the name of James Brindley who,

The derelict canalside Mather Lane Mill of 1882 in Leigh has an early concrete floor.

along with John Gilbert, engineered the canal. Scheduled passenger services began in 1776.

The Stretford & Leigh Branch of the canal was subsequently extended by the Leigh Branch of the Leeds & Liverpool Canal, which took it on from Leigh.

The limit of the Bridgewater Canal is marked by a stop board that is suspended from a manually operated iron and timber crane. This is one of a number that were installed to protect the 50km of water on one level.

Leigh was a coalmining town and also has a number of mills, with prominent chimneys, dating from Edwardian times. Frankie & Benny's and a canyon of houses are among the more recent developments by the canal. After crossing the line of a disused railway, the canal passes Butt's Basin with its wharf and canal office.

A new brick bridge has been built in a style in keeping with the original brick-arched bridges, and is followed by a school on the left, approached along the towpath by many pupils. The housing is a mixture of modern estates backing on to the canal and new terraces with roads meeting the canal at right angles.

A bridge gives one of a number of reminders of the 6km/h speed limit with another stop board crane close by. Glimpses of more open country appear with lower land leading on to Chat Moss, away to the south, and views northwards to the Pennines and a prominent aerial.

Astley Green is a colliery village, dominated by its 30m pithead gear, the last in Lancashire, and one of the largest steam winding engines ever built, the disused mine being used for the Astley Green Colliery Museum. A red-brick Victorian church stands to the south side of the canal with the waterside Old Boathouse public house built opposite.

The canal forms a southerly loop as it follows the contours. At intervals a blackened but delicate church spire stabs the sky above the housing. Up the hill from the one at Worsley is Old Hall, a 16th century manor house once owned by the Duke of Bridgewater. Extensively remodelled in the 19th century, it now has restaurants and hosts Jacobean banquets. Briefly back to the current day, as it is overlain by Junction 13 of the M60, the canal passes unexpectedly into one of the jewels of the Canal & River Trust network, the basin at Worsley and its surroundings. The canal goes into cutting, passing below a stop board crane and heavily

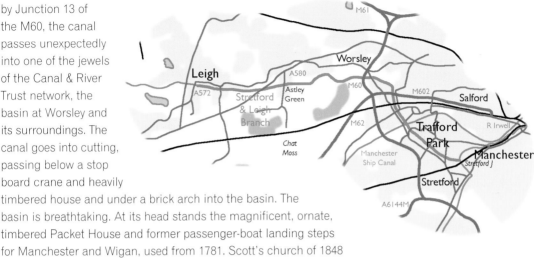

timbered house and under a brick arch into the basin. The basin is breathtaking. At its head stands the magnificent, ornate, timbered Packet House and former passenger-boat landing steps for Manchester and Wigan, used from 1781. Scott's church of 1848 contains a rich collection of Duke of Bridgewater monuments.

From the north-east corner of the basin leads a channel that divides before the Tung Fong restaurant. The channels pass under the A572: the one on the

The magnificent Packet House and passenger-boat landing steps in the basin at Worsley.

left through rough-hewn arches festooned with ivy, the one on the right through brick arches with interesting holes going off at angles. The two routes rejoin in a pond surrounded by a rock cliff. At the back are entrances in the rock. A double-track inclined plane operated between 1797 and 1822, protected at the top by locks. This served the mines, which were worked from the 14th century onwards and in which the Duke of Bridgewater cut 74km of canal tunnel on four levels. Despite the fact that this was the first of the industrial canals built over two centuries ago, those tunnels still account for 52 per cent of the canal tunnelling in Britain and exceed the total number of canal tunnels in the rest of the world. The mines are now considered too dangerous to enter but water is pumped out and the iron ore in it has coloured the canal orange for a considerable distance. There are hopes that some of the mined area can be made safe for public access.

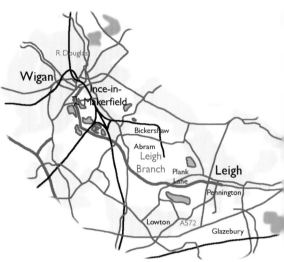

The southern end of the canal basin is flanked by wooded lawns and backed by a village store and the Bridgewater Hotel. It leads on to Worsley dry docks (perhaps the country's first, now used for work on narrowboats), Worsley warehouse (now restored for alternative use) and housing around the Worsley coke ovens' site. In the middle, the Sea Cadets have their TS *Ilex*.

As rapidly as the canal improved, it deteriorates with debris of all sorts over the next reach. There is housing following the right bank although the left is open parkland, once crossed by a railway passing close to the canal at Monton. The Crompton's at the Waterside bar and restaurant is situated at this point. A busy road follows the left bank until the M602 crosses both. The reach passes under the Liverpool to Manchester railway, after passing new housing. Another main road follows the right bank from the Wellington Inn through to Barton upon Irwell. By the A57 crossroads in Patricroft, the bank was landscaped. In 2014 thieves dug a 15m tunnel from the canal

bank to Tesco Extra, where they took £60,000 from the cash machine. Local amenities include the Packet House. The Dutton Arms is situated beyond a canal basin and there is an interesting set of warehouses.

Several tower blocks are passed before the run down to the Barton Swing Aqueduct, another wonder of the canal system. Originally a 183m x 12m clearance stone aqueduct crossed the River Irwell in three arches. Britain's first navigable aqueduct, it lasted from 1761 to 1893 when the Manchester Ship Canal was built. The original plan was to lock down and up again, then it was considered as a boat lift to clear the Manchester Ship Canal. The present solution was developed by Sir Edward Leader-Williams. Situated with the pivot point on an island in the middle of the Manchester Ship Canal and a tall brick tower at the west end, it has control huts on the two ends of the Bridgewater Canal, which operate stop gates. The girder swing aqueduct has a 72m x 5.8m x 2.1m deep tank that is swung full to save time draining. With 23m clearance over the Manchester Ship Canal, it weighs 1,600t and is opened half an hour before a ship is due on the major waterway in order to give the ship time to stop in case of a malfunction, any collision having potentially catastrophic consequences.

A pair of hotel boats cross the massive Barton swing aqueduct carrying the Bridgewater Canal over the Manchester Ship Canal.

Perhaps it is not surprising that the authorities do not rush to move this vast structure to let a small boat across. However, the walking route is not easy: a visitor must pass down a steep bank and through a narrow gate that leads directly on to the main road, then along the footway of the adjacent swing road bridge where there is an inevitable crosswind. The route then finally goes along a narrow footpath that is hemmed in between a fence and a hedge before there is a significant drop from sandstone blocks with rounded edges. Passed on the walk is the richly decorated Catholic church of 1867 by Pugin.

The character and setting of the canal now changes as it passes into the huge Trafford Park industrial complex. Built on land that had been owned by the de Trafford family since the time of Cnut, it was sold in 1896. Kelloggs' largest European factory was a major user of the canal until 1974. New construction is still taking place on the site.

The branch arrives at Stretford Junction or Water's Meeting and the main line of the Bridgewater Canal, part of the Cheshire Ring, with another Liverpool to Manchester railway line crossing on the far side.

Rochdale Canal

During the era of canal mania it took just an hour in 1791 to raise
£60,000 in subscriptions to build the first trans-Pennine canal.
Calderdale was the centre of the British wool industry in the 17th and
18th centuries and completion of the Rochdale Canal in 1804 provided
a much-improved link with Manchester. Surveyed by John Rennie, it
was built by William Jessop, who
didn't like tunnelling, preferring
to add extra locks to climb over
obstacles. This is the only trans-
Pennine canal without a summit

*Tuel Lane Tunnel
at Sowerby Bridge.*

tunnel, offering splendid scenery, particularly on the West Yorkshire side of the summit, but producing a large number of locks: 36 to the summit and 56 down to Manchester, a poem having been commissioned for each lock. It was built as a wide-beam canal, passing its last working boat in 1937. Apart from the Cheshire Ring section, it was abandoned in 1952 but remained as a water supply feeder, requiring structures to be maintained. Since restoration, there have been several serious breaches.

After the lock in front of the William IV is the Tuel Lane Tunnel. The 100m long tunnel replaces a section filled in to carry the A58 and a road off it. Reopened in 1996, this is one of the most complicated examples of canal restoration work. The tunnel curves at the end and is subject to considerable turbulence from the lock. It may not be entered without the lock keeper's permission. Locks on the canal were set at a 2.7–3.0m rise to allow interchange of gates. The Tuel Lane Lock replaces two former locks and, at 5.94m deep, is now the deepest inland lock in the country. The walking route is past the Lock Keepers Tavern.

Climbing beyond Friendly, the A646 follows the canal and river, as does the Leeds to Manchester railway. This line, engineered by George Stephenson, was the world's second to carry both freight and passengers and is usually within a kilometre of the canal or much less. There was a station at Luddenden Foot, where Bramwell Brontë, less well known than his literary sisters, was briefly a clerk until he was sacked for being drunk. He used to drink in the Lord Nelson. Watering holes near the canal are the Old Brandy Wine and the Coach & Horses.

Mytholmroyd is sited where Cragg Brook joins the River Calder and was home to coin counterfeiters. David Hartley, leader of the Cragg Vale Coiners, was hanged in 1770 and buried in Heptonstall. It was the birthplace of former Poet Laureate Ted Hughes and is renowned as the venue for the world dock pudding championships. Beyond the White Lion Hotel, the A646 crosses, to cross back over the recent Falling Royd tunnel, a structure made of corrugated arches on a curved line through which daylight is not visible; it is surprisingly dark inside for its large size.

One of the stone mills houses a selection of shops in Victorian style, another has a Thai restaurant and there is an alternative technology centre. Hebden Bridge is a town developed on the manufacture of fustian, inspiration for John Morrison's Milltown books. Hebden Bridge Vintage Weekend, in early August, follows the town's Arts Festival. A small but open marina is sited in front of the Railway Hotel and Hebden Lodge Hotel.

Distance
26km from the Calder & Hebble Navigation to the Bridgewater Canal
Highlights
Great Wall of Todmorden and Gauxholme Viaduct
Navigation Authority
Rochdale Canal Company
OS 1:50,000 Sheets
103 Blackburn & Burnley
104 Leeds & Bradford
109 Manchester

Beyond the Little Theatre, a four-arch aqueduct carries the canal over the River Calder, just downstream of its confluence with Hebden Water, which carves a deep valley down from the moors and passes under a stone arch bridge of 1510, replacing a medieval timber bridge. The 17th century building of Heptonstall Grammar School, featuring a museum with farming, crafts and handloom weaving, overlooks the valley. Also of interest is the chapel of 1764, which was used by John Wesley and was one of the first octagonal Methodist chapels.

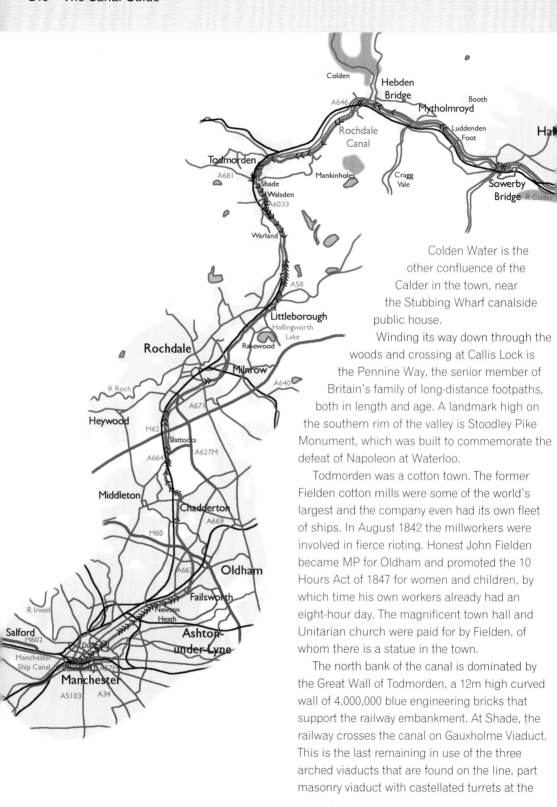

Colden Water is the other confluence of the Calder in the town, near the Stubbing Wharf canalside public house.

Winding its way down through the woods and crossing at Callis Lock is the Pennine Way, the senior member of Britain's family of long-distance footpaths, both in length and age. A landmark high on the southern rim of the valley is Stoodley Pike Monument, which was built to commemorate the defeat of Napoleon at Waterloo.

Todmorden was a cotton town. The former Fielden cotton mills were some of the world's largest and the company even had its own fleet of ships. In August 1842 the millworkers were involved in fierce rioting. Honest John Fielden became MP for Oldham and promoted the 10 Hours Act of 1847 for women and children, by which time his own workers already had an eight-hour day. The magnificent town hall and Unitarian church were paid for by Fielden, of whom there is a statue in the town.

The north bank of the canal is dominated by the Great Wall of Todmorden, a 12m high curved wall of 4,000,000 blue engineering bricks that support the railway embankment. At Shade, the railway crosses the canal on Gauxholme Viaduct. This is the last remaining in use of the three arched viaducts that are found on the line, part masonry viaduct with castellated turrets at the ends and more hidden girders added in 1906, which now carry the live load.

Travis Mill Lock lifts the canal past the Cross Keys. A succession of further locks continue the process. Near Bottomley Lock, Stephenson gave up the contest and took the railway into a 2.6km tunnel, the longest railway tunnel in the world when he built it in 1840.

A back-pumping scheme has been installed at Warland Lower Lock to assist water levels at the summit. Warland Upper Lock, in front of the Bird i'th Hand, is the boundary between West Yorkshire and Greater Manchester.

The summit level is 183m above sea level, Britain's second-highest, and those who get this far can buy summit plaques at Longlees Lock, in aid of the canal society. On the road, Todmorden Turnpike tollhouse at Steanor Bottom Bar still lists the toll charges. An inclined trackway ran down to the wharf from a quarry and brickworks. The Summit public house has a backdrop of a high stone quarry face at the foot of Blackstone Edge, on which were built Warland and Light Hazzles Reservoirs as canal feeders.

Beyond the A671 is the end of the Rochdale Branch, which ran northwards for a kilometre towards the centre of the town. Rochdale – named after the River Roch, in turn named after the British Celtic *rached*, river by the forest – was a cotton town and, like Todmorden, its millworkers were involved in fierce rioting in August 1842. Two years later 28 weavers set up the original Co-op store in Toad Lane (t' owd lane), a scheme that worked where others had failed because it bought in bulk, sold at market prices and shared the profits in proportion to purchases. By 1915, 3,000,000 people were claiming their dividends from the Co-operative Wholesale Society. The Rochdale Pioneers Museum is located in the original store. The town hall is Victorian Gothic with fine stained-glass windows, paintings, carvings, ceramics and a hammer-beam roof; it is one of the finest in the country.

The M62 trans-Pennine motorway is taken over a bridge that provided access to a farm and now takes the canal and a floating towpath, to the farmer's annoyance. There were a series of

Rhododendrons and dense greenery at Eastwood, hardly typical Pennine scenery.

problems in the area, including breaches and attempted breaches of the canal embankment, and all the balance beams were sawn off the lock by the M62. Eventually, the farmer was found to be responsible and jailed in 2008.

As the traffic roar falls away there is not a building in sight, just golfers up on a ridge to the right. Powerlines follow the canal to Chadderton and the six Laneside or Slattocks Locks are crossed by the A664 and the railway again. To the west is Middleton, with the Bobby Charlton Soccer School Academy.

Chadderton is the haunt of the Broadway Boggart, identified on misty nights by flickering blue lights in the sky and disturbing crackling and screeching noises.

Blocks around Piccadilly Lock include County Hall and the glass-and-concrete Piccadilly station, which began life as Store Street and

Moorings alongside an old mill at Eastwood.

Trees give way to moorland near Lumbutts.

became London Road in 1846. In 1881 it had a major rebuild with four spans of 24–30m arches on lattice girders supported on cast-iron columns. Its 1.2ha undercroft is an abandoned warehouse with heavy brick arches and cast-iron columns, the spaces between the columns being considered ideal as a tram interchange for the Greater Manchester Metrolink.

This is one of the finest Victorian city centres anywhere, Manchester having been built on textile wealth that developed from the skills of Flemish weavers in the 14th century. Although the final section of canal through Manchester is largely below road level, the whole area has been restored and sees a lot of city centre activity.

Chorlton Street lock keeper's cottage is built over the canal. Italianate buildings are the old Minshull Street magistrates' court. Manchester Metropolitan University is in an 1895–1912 French Renaissance-style listed building. The North West Film Archive has material from 1896 onwards. The Grade II Mechanics' Institute was the first meeting place of the Trades Union Congress in 1868, and the Co-operative Insurance Society and the University of Manchester Institute of Science & Technology were also founded here. Manchester Art Gallery has one of the country's best collections of pre-Raphaelite art.

After Oxford Road Lock comes the line of the Manchester & Salford Junction Canal, built to link the Rochdale Canal with the River Irwell, its tunnel being used as an air raid shelter during the Second World War. It opens into a

Gauxholme Viaduct on the world's second-oldest passenger and freight railway line. They don't build them like that anymore.

basin below the glass edifice of the Bridgewater Concert Hall, opened in 1997 as the new home of the Hallé Orchestra. Beyond is the Manchester Central Convention Complex in the former Manchester Central station railway terminus, a station with a roof span exceeded only by that of St Pancras in London. Built in 1880, the 64m arch rises 27m to provide an extensive unobstructed exhibition hall.

Tib Lock has the Canal Bar alongside with plenty of the customary broken glass and the A5103 passing over a tunnel. The Manchester Sound derived from the Hacienda Club and featured such names as Take That, Oasis and The Stone Roses. It is now apartments. What at first glance appear to be four wharf cranes on the side of a modern building metamorphose into four griffin-like sculptures with chains in their mouths.

Many Roman remains have been found around the area and there would probably have been more if the fort of 79 had not been largely destroyed by the construction of the 200-arch 19th century railway viaduct, which now carries the Manchester to Northwich and Metrolink lines by the canal. The Dukes 92 public house takes its name and number from Duke's Lock, the last on the canal, and is located in the stables of the Merchants Warehouse of 1827, the oldest surviving great warehouse.

The Rochdale Canal joins the Bridgewater Canal as canals lead off in all directions and a network of railway and footbridges provide a hive of transport activity.

A brick mill at Miles Platting starts the build-up to central Manchester.

35 Ashton Canal

Despite its connections, the Ashton Canal, designed by Thomas Brown, was built as an isolated narrow canal from Ashton-under-Lyne to Manchester. It was opened to Ancoats in 1796, extended to Ducie Street in 1799 and connected to the Rochdale Canal the following year. Initially it was used to take fresh vegetables from the market gardens of Dukinfield into

The Peak Forest Canal joins at Dukinfield Junction.

Manchester and later it served many small coal mines in the area. Commercial traffic ceased in 1957 with the last pleasure boat in 1961. Restoration began in 1968. Full restoration also preserved the Cheshire Ring, including the lower end of the Rochdale Canal. There is now considerable use by narrowboats.

Ashton-under-Lyne was a small market town and, indeed, still has a thriving outdoor market most days. With local coal and goods transport by canal and, subsequently, railway, it became an important cotton centre. It saw the start of the Chartist action in 1842 when strikers turned workers out of the mills.

The Peak Forest Canal also brought Macclesfield Canal traffic – coal and lime for industry, building and farming – approaching across an aqueduct over the River Tame and under a slender, stone towpath bridge for the Ashton Canal at Dukinfield Junction. The aqueduct was built by the promoters of the Ashton Canal in readiness for the Peak Forest Canal.

Opposite the junction was built the New Ashton Warehouse in 1834, to serve the growing number of mills on the west side of Ashton-under-Lyne. All but the ground floor was destroyed in a fire in 1972. The remains have been rebuilt as the Portland Basin industrial museum with a 1920s street. A prize possession outside is the 7.3m diameter breast shot waterwheel, which was used to power the cast-iron hoists.

A footbridge with a handrail missing carries the towpath over the entrance to the former Princess Dock, constructed for easier transhipment between railway and canal, although this worked to the railway's advantage. Brookside Sidings now occupy the dock site for 1.2km, flanked by the Huddersfield to Manchester railway line. This is joined just before Guide Bridge by the line from Glossop.

Oxford Mills, opposite the sidings, were built 1845–1851 by the Mason family for

Distance
11km from the Huddersfield Narrow Canal to Ducie Street Junction
Highlights
Velodrome, squash centre and Etihad Stadium
Navigation Authority
Canal & River Trust
OS 1:50,000 Sheet
109 Manchester

spinning. They were supplied with housing and community facilities, including a sports ground, swimming and washing baths, an institute and library, plus smoking and chess rooms – an advanced concept for the time.

Guide Bridge takes its name from the bridge over the canal and its signpost to Ashton, Manchester and Stockport. Since it acquired its name, the bridge has been widened to an 85m tunnel although the original bridge is still obvious from water level. Guide Bridge Mill on the right was a late-Victorian cotton-spinning mill.

The Ashton Packet Boat Company Ltd basin is the site of narrowboat restoration and horse-drawn canal trips. To the west is another freight line and, over that, a road bridge with some intricate brickwork under its high, skewed arch. The gap in the buildings has been used for the M60 Manchester Outer Ring Road, near Audenshaw Reservoirs.

Park House offers some of the most attractive accommodation on the canal, sheltered housing established in 1969 by Air Vice Marshal Johnnie

Working through the top lock at Fairfield Junction, one of three formerly double locks. Beyond the lock, members of a canoeing class prepare to go afloat at the Water Adventure Centre, which does much for disadvantaged youngsters.

Johnson. Behind the new housing on the other bank is Fairfield, a 200-year-old Moravian Settlement village.

Fairfield Junction brings the stump of the Hollinwood Branch. This climbed away through Droylsden to Oldham. It was abandoned in 1932 but the first 160m is now being restored as a marina. Another section remains in the Daisy Nook Country Park. There was also a Fairbottom Branch but this has now disappeared.

Now begins the series of 18 locks descending to Piccadilly. The locks are in good condition although the use of scaffolding-style safety rails owes more to economy than to beauty. Many of the locks have overflow channels and about half of these are on the verge of being shootable, subject to sufficient water flow and boats of the right dimensions and materials. The first lock was subject to an arson attack in 1961, disrupting an Inland Waterways Association protest rally in Manchester.

The first two locks were doubled in 1830 but only the left one of each pair has been retained in use. Between them is a packet boathouse of 1833, which had services to Ashton, Stockport, Stalybridge, Hyde, Marple and Manchester. The services were regular and efficient and Ashton to Manchester took only two and a half hours, including working through the locks. The building is now part of the Water Adventure Centre.

Locks 16 to 8 form the Clayton Flight over the next 1.6km, several with distinctive characters. Lock 13, Clayton Top Lock, is overlooked by the Strawberry Duck public house. Between locks 11 and 10, Britain's deepest narrow lock at 4.22m, is the bridge carrying the towpath at Clayton Junction over the Stockport Branch. This ran for 8km, serving much industry. It was not used from the 1930s and was closed in 1960.

Standing back from lock 9 and the Bridge Inn is the church of St Cross with St Paul. It was built in polychrome style in 1866 by William Butterfield and is one of three Grade I churches in Manchester.

A large girder bridge once carried a freight railway but now has only bushes on it. The new A6010 bridge crosses near where the Clayton

ranch served a complex of chemical works
until it was closed in 1953. Before it on the
right is the National Cycling Centre of 1994 with
Britain's first indoor Olympic track and 12km of
mountain bike trails. It is followed on the right by
the Indoor Tennis Centre and on the left by the
National Squash Centre and the Etihad Stadium
of Manchester City, the Manchester football
club supported by Mancunians rather than
the one supported by the rest of the world. The
concentration of sports facilities in SportCity is a
2002 Commonwealth Games' legacy.

Locks 7 to 4 form the Beswick Flight over 500m.
The first has a lock keeper's cottage of 1865. At
this point the Medlock Valley Way footpath joins
the towpath and faces the turn at the west end of
the lock. This is very constricted. Between here
and the next lock, Bradford Lock, the Bradford
Branch on the right served the Bradford Colliery.
Both the branch and the Colliery have now gone.
The Beswick Locks are the last two of the flight.

An overflow weir disappears under a concrete-
plank fence to join the River Medlock, which is
underground at this point. The river resurfaces at
the Medlock Aqueduct, an arrangement that is
made more complicated by a large railway bridge
crossing over the top of it and a solid brick wall

reaching from the railway bridge to the towpath
on the right.

The final three Ancoats Locks come in
200m. There were several small branches, the
largest being the 400m Islington Branch. This
left Ancoats Junction on the right, running
past the heavily fortified Canal & River Trust
office, to supply coal, salt and sand wharves.
An arm opposite served the Manure Wharf,
which supplied night soil and manure from the
corporation stables to the Beswick Sanitary
Works and farmers further along the canal.

Ducie Street Basin had a massive warehouse
for storing high-quality salt from Northwich but it
was burned down a few years ago. Still present
is the impressive London Warehouse, a railway
warehouse built in 1867 with wrought- and cast-
iron to reduce fire risk.

The Ashton Canal joins the Rochdale Canal at
Ducie Street Junction, alongside the Dale Street
car park. This occupies the filled-in Dale Street
Basin, overlooked by one remaining warehouse.
To the west, a lock continues the Cheshire Ring
through the centre of Manchester.

The massive brick London Warehouse. The
Rochdale Canal lies to the right under new offices in
contemporary styling.

Peak Forest Canal

The Peak Forest Canal must have the most inviting name of any of our canals. Anything it lacks in the way of forest these days is compensated for by peaks, particularly on the upper section.

Designed by Benjamin Outram, it was opened in 1800 to service the limestone quarries at Doveholes. Although the upper end of the canal is one of the highest parts of the British canal system still in operation, it did not reach all the way to the quarries. In 1799, the 10km horse-drawn Peak Forest Tramway was opened to connect the two. A vast complex of wharfs was constructed at Buxworth or Bugsworth, as it was called before the inhabitants changed the name. The wharves are located around the Navigation Inn. The area with its limekilns was handling 600t of limestone and lime daily in the 1880s. The tramway was closed in 1926.

The Whaley Bridge Arm provided a connection with the Cromford & High Peak Railway and thence to the River Trent.

The canal follows the Goyt valley, clinging to the left side and offering spectacular views out over it through not infrequent breaks in the dry stone

Just one part of the Buxworth canal terminal complex.

walls. The canal is popular. The towpath is much used by walkers, who appreciate the fine views.

The bridge carrying the A6015 at New Mills is heavily shored. Its courses of stonework show a marked dip above the canal. Road vehicles are reduced to single file in an effort to keep the two forms of traffic apart.

New Mills, inspiration for the New Mills clog dance, has had textile printing, engraving, a canal foundry and engineering industries. The largest mill is Swizzles Matlow's sweet factory and the smells emanating are likely to produce a craving for their Refreshers and Love Hearts. The canal user should plan ahead by carrying a suitable stock.

Fencing off an old stone quarry has produced a field of the kind that only goats can fully appreciate. The resident family have trodden paths in places that gravity should have placed out of reach. A copse follows, banks of wood sorrel sloping down to waterside marsh marigolds. These are followed by neatly mown lawns each side of the canal at Disley. Here are some of the best views of all. It is easy to understand why this has been the scene of two breaches, the first in 1940 and the other in 1973, which swept away four cruisers after the bank collapsed, owing to a prolonged spell of torrential rain.

The Ordnance Survey have placed a trig point just 200m from the canal at Marpleridge. Oak trees and yellow-flowered gorse bushes cover fields to the west of the canal. Breaks in an ivy-covered wall and a stand of pine trees allow a last look across the Goyt valley to Mellor Moor and beyond. This is excellent touring country.

By this stage the high land on the west is already subsiding and the canal leads out on to a peninsula, where it is joined by the Macclesfield Canal emerging under a roving bridge to become part of the Cheshire Ring, more enclosed by trees from here. The Macclesfield Canal did much to improve the prosperity of the Peak Forest Canal.

The Marple aqueduct drops to the River Goyt way below.

Distance
23km from Buxworth to Dukinfield Junction
Highlights
Marple lock flight and Possett Bridge
Marple Aqueduct and railway viaduct
Navigation Authority
Canal & River Trust
Canal Society
Marple Locks Heritage Society
www.marplelocks.org.uk
OS 1:50,000 Sheets
109 Manchester
110 Sheffield & Huddersfield

While the upper Peak Forest Canal has always been kept open, the lower section was abandoned in 1944 and reopened only in 1974.

Conveniently, all 16 locks on the Peak Forest Canal are in a single flight at Marple, which some consider to be the most agreeable urban flight in England. It was not completed until 1804, a tramway being used up and down the hill for the first four years of the canal's life.

For nearly a century from 1840 onwards, narrowboats were built and repaired at the top of the flight. The dry dock these days acts as an interesting sunken flower garden.

A road separates the locks from the onlooking row of houses on the left. On the right, long side ponds run out along the contours with modern houses between each one. After the fourth lock, Possett Bridge is a particularly fine structure, carrying the road across. It has a horseshoe-shaped arch on the left for tow horses and this leads out on to a paved area, surrounded by flowerbeds. There are interesting buildings all around the flight, from offices in the 1805 Oldknow's cotton warehouse to the quaint Brabyns No. 1 Cottage. The flight descends further through tree-covered fields, overlooking Brabyns Park. Three-quarters of the way down, the railway tunnels under the flight, seeming not to be deep enough to clear the bottom of a lock, especially as some locks here are close contenders for the title of deepest narrow canal lock. The locks are also unusual in having stone-arched bridges across their lower ends. Uphill walls are curved so that a boater, pushing the balance beam to open the gate, walks round from the side of the lock on to the bridge. The balance beams then prevent the bridge being used easily with the bottom gates open.

Almost immediately comes the magnificent Grade I Marple Aqueduct of 1801, its massive proportions only overshadowed by the nearby railway viaduct. The three arches of the aqueduct, designed with circular piercings through the piers below the trough, consumed 600m³ of masonry in their construction. It was repaired after a partial collapse. The water level is well over 30m from the River Goyt below, Britain's highest stone aqueduct with circular holes in the spandrels between the three 18m spans, supported on cylindrical piers.

Rose Hill cutting is equally narrow, with a high retaining wall on the right.

Butterhouse Green Tunnel is surrounded by greenery.

Its unusual shape results from the fact that it began life as a 100m long tunnel, not being opened out until 1820.

The canal dives into the 282m Hyde Bank Tunnel, a squat structure with the towpath running over the top and a slight kink in the middle, to emerge among wooded banks that are covered with celandines, bluebells and red campion in early summer.

Beyond Romiley, a high railway bridge of almost tunnel proportions crosses the canal. The first half has significantly smaller dimensions than the far end and the step must have caused the arches to ring with curses from southbound boats many times over the years.

Naylor's abrasive mill is the first of an increasing number of mills, many derelict. Some provide an atmosphere of decaying charm while others could benefit from the attentions of a bulldozer, an idea that is receiving some positive application around Hyde.

Captain Clark's Bridge, at Hyde, is a fine, stone roving bridge and is paired with another roving bridge with cast-iron deck to take the towpath across the canal. After passing a mill the slightly realigned canal is crossed by the M67 on a bridge that is unusual in having the luxury of lighting from underneath in the daytime.

37 Huddesfield Narrow Canal

The Huddersfield Narrow Canal was designed by Benjamin Outram as the most southerly and shortest of the three canal links across the Pennines. Begun in 1794, it was opened in 1811. It was not a great success commercially. Joining the eastern end of the Ashton Canal in Ashton-under-Lyne at its first of 74 locks, it had a direct connection to the Cheshire Ring, allowing access to Lancashire, Merseyside and the Black Country. Closure came in 1944 but it was retained as a canal water supply and reopened in 2001.

Approaching the prominent radio aerial above Grasscroft, the canal crosses over the river on small aqueduct. The placid cricket ground nearby on the right has a panoramic backdrop of crags and scars leading up to Saddleworth Moor, the Peak District National Park and an obelisk above Tunstead to watch if the game gets tedious. After Saddleworth Rangers Rugby League Football Club ground, the athletic activity continues with a running track and a sports field the other side of the A669. After the A670, the canal flanks a landscaped region at Uppermill.

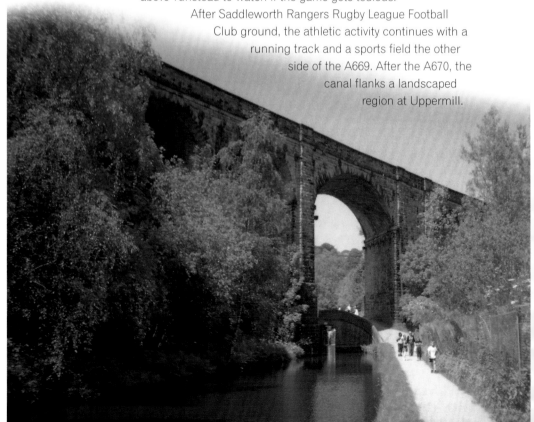

A grassed area and car park are surrounded by the Granby Arms, the Waggon Inn and Saddleworth Museum and art gallery in a former cotton mill with a power loom and clothier's cottage.

So far, the locks have been surprisingly well spaced. From Dobcross to Diggle they come as a flight, climbing up on to more exposed moorland above the treeline. From the top lock the water level is 196m, the highest connected canal level in Britain. As the canal approaches Diggle, set in a basin of hills, the railway closes alongside and suddenly the tunnel mouth appears ahead – an archway that seems to be set too far forward from the hillside – an entrance to one of the wonders of the canal world. The original tunnel mouth had to be moved north in 1849 to accommodate the railway alongside. Until entry is permitted, the tunnel mouth is closed with a gate showing a life-sized narrowboat bow with a legger on each side. This is Standedge tunnel. At 4.95km, later extended by 250m, this is the longest and highest canal tunnel in the UK, burrowing under the Pennines at depths to 194m. Opened in 1811, it took 17 years to build. The 2011 record for legging through is under two hours.

Information boards have been erected beside the canal near Marsden. More entertaining than informative is the revival in the 1990s of the Celtic Imbolc cross-quarter festival, in early February, with fireworks, drummers, live music, theatre and a torch-lit procession near the canal.

The start of the long flight down soon arrives with the Railway pub facing its namesake across one of the first locks. Also crossing here is the stiff 20km Colne Valley Circular Walk, which re-crosses at Golcar.

Approaching Slaithwaite, the River Colne is never far away on the right and sometimes exposes bare crags. These enhance the scenery; steep hillsides with farms, weavers' cottages and,

The canal passes under Saddleworth Viaduct at Uppermill.

> **Distance**
> 37km from the Ashton Canal to the Calder & Hebble Navigation
> **Highlights**
> Standedge Tunnel and visitor centre
> Locomotive Bridge in Turnbridge, with a windlass-operated vertically lifting deck
> **Navigation Authority**
> Canal & River Trust
> **Canal Society**
> Huddersfield Canal Society
> www.huddersfieldcanal.com
> **OS 1:50,000 Sheets**
> 109 Manchester
> 110 Sheffield & Huddersfield

increasingly, mills, all in darkened stone.

Slaithwaite is steeped in history. Slaithwaite Manor is a Grade II Elizabethan house, dating from the 1560s. The nearby St James' church was built in 1789 to replace one that was subject to flooding by the River Colne. Its shape is based on the Puritan ideal that it was easier to be devout in a barn than in a temple. Its name was twisted into Sanjimis, a four-day festival of feasting, dancing bears and Waffen Fuffen bands (dialect for waifs and strays) playing in the contemporary equivalents of jug bands. The present funeral parlour was the old Slaithwaite Free School, a fine building. The canal passes Empire Brewing, public toilets, the Shoulder of Mutton and the Commercial before moving behind the Globe Worsted Company and the Shaw Pallet factory.

From 1825, Slaithwaite was a spa based on mineral springs in the bed of the River Colne. It modelled itself on Harrogate. The village was a hive of industry. Smuggling was rife in the 19th century too; the usual procedure was to place things in the canal by day and then hook them out by night. One gang, caught in the act by the king's men, pretended to be drunk and claimed

The canal runs right through the centre of Slaithwaite, once again.

to be trying to rake the moon's reflection out of the canal, thus acquiring the title of Slawit Moonrakers.

The route becomes rural once more. It is progressively more overlooked by houses on the A62 at Linthwaite and Milnsbridge. Westwood Mill is the oldest in the valley, on what was a fulling mill site in 1604. Old mill buildings have been fitted with incongruous balconies and now serve as residential buildings, accompanied by a tall chimney. Three 19th century cottages in Golcar form the Colne Valley Museum, including an 1850 weaver's living room.

The Luddites were active in Milnsbridge the year after the canal was opened. They shot a mill

The distinctive Locomotive Bridge in Turnbridge.

owner from Marsden, broke into a number of mills and smashed up cropping frames, which they thought would put them out of a job.

The Huddersfield Broad Canal, also known as Cooper Canal or Sir John Ramsden's Canal, after its promoter, continues northwards. It was completed in 1776 in order to bring boats off the River Calder with coal for the mills, carrying commercial traffic until 1953.

Huddersfield is an industrial town developed around its woollen mills and is unashamedly Victorian. St Peter's church of 1838 and the 1902 clock tower are both Victorian Gothic while the station of 1847 by JP Pritchett is Corinthian, one of Britain's finest railway buildings. The Lawrence Batley Theatre is in a former Wesleyan chapel and the Huddersfield Choral Society are based in the 19th century town hall. Huddersfield Art Gallery offers Gainsborough, Lowry, Turner and Constable with 20th century British art and there is a Huddersfield Contemporary Music Festival. The town was one of the first places to have a council housing scheme and produced Harold Wilson. There is even a town crier.

At the foot of a stone mill chimney in Turnbridge, the Locomotive Bridge of 1865 is an unusual survivor; it is a bridge with a deck that is lifted vertically by windlass. It leads down towards the Galpharm Stadium, occupied by Huddersfield Town Football Club and Huddersfield Giants Rugby League Club, a golf driving range and a multiplex cinema. The A62 crosses by a canal mural, the former tram depot and the Automobilia Transport Museum.

The locks begin with Fartown Green Lock at Red Doles and playing fields begin on the east side of the canal. Falls Lock is by Trafalgar Mills, a former worsted mill, and the remains of Fieldhouse clay works accompany Fieldhouse Green Lock. Riddings Lock, near Kirklees dog pound, has cranked balance beams as the adjacent bridge has been built very close and it has guide pins for ropes, as have some other locks in this flight. Turnpike Road or Halewood Lock is near a former cornmill, the A62 Leeds Road, to which the name applies, crossing on an extended stone arch. A sports field notice has warned about the danger of javelins but not what to do if one is thrown in your direction. Beyond, North Moor provides an impressive backdrop to Kirkheaton.

The following railway viaduct across the canal was intended to carry nothing less than the main line between England and Scotland. The canal arch collapsed during construction in 1866. Its yellow colour contrasts with the dark blue used for the rest of the structure. In practice, it had to settle for the Kirkburton Dick shuttle service.

Longlands or Vernon's Lock at Deighton is followed by Ladgrave or Johnson's Lock, the latter with a very similar layout to Riddings. Another disused railway viaduct follows, again in blue brick but with 15 arches this time. Completed in 1910, it saw only 27 years of use.

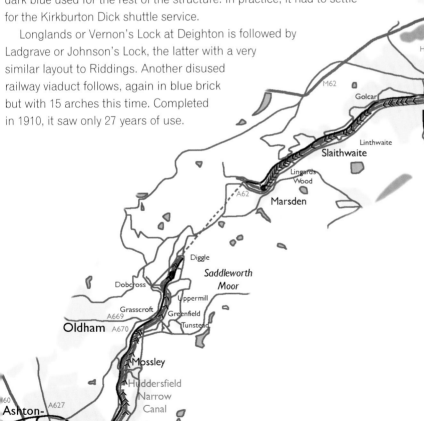

38 Calder & Hebble Navigation

The construction of the navigation parallel to the River Calder and, in places, using its course, was completed in 1770. This was one of the canals that made a major contribution to the Industrial Revolution, linking the Aire & Calder Navigation with the Rochdale Canal and providing a route from industrial Yorkshire over the Pennines. At no point is it far from conurbations yet it is surrounded by far more open country than the map might suggest.

Its western terminus is the basin at Sowerby Bridge, deep in the valley of the River Calder. Hillsides rise steeply to north and south. Roads, some still cobbled, are equally steep as they climb away from the grimy industrial buildings. The canal basin is a place where new life has come to the old buildings in various forms. The Bolton Brow Gallery featured in the BBC's *True Tilda* series and the Moorings bar is in an old canal building.

Just beyond the canal basin, the Rochdale Canal leaves on the right, opposite the first of the Navigation public houses.

Assorted industrial smells permeate the air at times all along the canal yet some industry has been replaced by housing and trees gradually replace the buildings. The open hills rise behind. Railways and roads intertwine around the navigation in a three-dimensional tangle.

Looking east from the Sowerby Bridge basin.

The railway crosses for the first time at Copley, the viaduct passing over both canal and river.

Along the entire length of the navigation, signposting is clear. The first warning comes at Salterhebble, where the remains of the Halifax Branch joins. This branch allowed Halifax, founded on the cloth industry, to be built back from the main line of the canal. One of the nearest features is the Perpendicular St John the Evangelist church, with its pinnacles, parapets, gargoyles and fine woodwork.

Salterhebble brings the end of the top pound, with a flight of three locks that replace a former staircase, wasteful of water, and the whiff of an adjacent sewage works. When the water authorities were created in 1974, the new Yorkshire Water Authority acquired this large works, which was the most neglected major one in the country. At that time, the effluent going into the Calder was more polluted than when it arrived at the works.

A stream passes under the navigation before the third lock. Beyond the lock is a modern, electrically operated guillotine gate. There is a small horseshoe-shaped tunnel under the road towards the Calder & Hebble public house.

The next lock, Long Lee, brings the dark River Calder alongside for the first time. At Elland, square, tapered chimneys with iron banding add a further dimension to the industrial architecture. A warehouse, which had large end doors that opened to admit a narrowboat, has become a private dwelling. Opposite, the Barge & Barrel has a welcoming notice for boat people.

There are 39 locks, of which 12 are flood locks. These may be left open when the River Calder is not in spate. Lock approach walls are often quite high. The locks themselves are unusual in having open panels to allow excess water to weir over them and by being operated by a bar of wood used as a crowbar rather than the conventional windlass. Substantial rope guide pins mark the corners of locks.

Distance
35km from Sowerby Bridge to Wakefield
Highlights
Wakefield's 99-arch railway viaduct
Navigation Authority
Canal & River Trust
OS 1:50,000 Sheets
104 Leeds & Bradford
110 Sheffield & Huddersfield
(111 Sheffield & Doncaster)

The Colliers Arms is handily close to the navigation while a sign for the Rawson Arms points vaguely in the direction of a wooded bank.

Beyond Cromwell Bottom, the navigation passes between lakes that look man-made and the waterski jump on one is an indication that they are well used. The Casa del Lago stands facing the navigation across one of the lakes in a pleasantly rural setting. This disappears at the next corner, just as the navigation crosses Red Beck at surface level. It passes between Brighouse and Rastrick, names evocative of coal mining and brass bands. The flowerbeds of the Black Swan, the Atlas Mill brewery tap, come down to the waterside, a foretaste of the landscaping that has gone into the canal basin at Brighouse. An astonishing double roving

A viaduct carries the railway over for the first time at Copley.

bridge at the start of the basin seems to offer towing horse lines excessive options, especially as the towpath does not continue beyond the basin. Indeed, the towpath is present only on the sections of cut but not alongside the natural river. Signposting for walkers is poor.

Just before the M62, the canal cuts right, avoiding two weirs. The country opens out again and the 330m Grade II concrete television mast on Emley Moor is seen for the first time, the UK's tallest freestanding structure. This is a replacement for one that collapsed in 1969 when ice brought down the guy cables.

After rejoining the river, the supposed burial place of Robin Hood is passed at the top of the

The River Calder sections are of a larger scale and lack a towpath, or sometimes any path at all.

wooded hill on the left. Although it might not be a particularly handy spot for Sherwood Forest, this is not as unlikely as it might at first seem as the grounds are those of the 16th/17th century Kirklees Hall, formerly those of the Cistercian Kirklees Priory, of which Robin Hood's cousin was prioress.

A canal leg runs past a wharf complete with original wharf crane. A river leg ensues before the route returns to the canal at Mirfield. A church is by George Gilbert Scott. Charlotte Brontë was at school here and the mills feature in her *Shirley*.

Barges built at Mirfield for the navigation have been based upon Humber keels, resulting in the short wide-beam locks. Lack of space has resulted in some spectacular sideways launchings over the years.

After the Navigation public house a couple of quaintly named works sit adjacent to each other, the mill of Squire A Radcliffe and Sons Ltd and the premises of the British Bung Manufacturing Company.

A confusing part of the navigation is by the Swan at Shepley Bridge Lock, where two sets of lock gates face the boater. The left set guard a barge dry dock, the navigation running right to rejoin the river opposite the Ship Inn. One more brief canal section comes before Ravensthorpe. A warehouse displays gantries and a wharf crane.

The Perseverance public house lies in a bleak partly derelict area. The Dewsbury Branch leads to the Savile Town Basin, serving this old industrial and heavy woollen-manufacturing town.

The Figure of Three Locks at Healey now lack the third lock, which once connected with the Calder.

A prominent and striking church in Horbury was finished in 1791, built and paid for by its architect, John Carr. A subsequent vicar, Sabine Baring-Gould, composed *Onward Christian Soldiers* for his younger congregation members to sing while walking to and from church.

A railway bridge over the river is continued as a 95-arch viaduct, the full expanse of which is not seen from the river. It is possible to see the 15th century Perpendicular cathedral, an uprated parish church.

At the Canal & River Trust depot, the final cut runs past a timberyard and some larger moored boats that are used as houseboats. The final lock, Fall Ing Lock, takes the navigation down to the River Calder for the last time.

Canalside mills near Rastrick.

39 Leeds & Liverpool Canal

Work started on the Leeds & Liverpool Canal in 1770. By 1777 it had been completed from Leeds to Skipton and from Wigan to Liverpool. It then became delayed by technical problems and expenditure on the Napoleonic Wars and it was not until 1816 that John Longbotham managed to complete the longest single canal in the country and the one that took the longest time to build. Once built, it was one of the most prosperous, especially from 1820 to 1850, promoting work in the mill towns and carrying coal, limestone, cement, machinery, wool, cotton, groceries, beer and spirits. Liverpool corporation had helped with the financing. Most traffic was at the two ends rather than across the Pennines but the finances were sufficiently healthy to support the whole canal and it has been the only trans-Pennine canal to remain open throughout. From the 1870s it lost longhaul traffic to the railways, the decline being exacerbated by water supply problems, but it was the 13-week freeze in 1963 that finished much of the local traffic.

It is perhaps England's finest canal for scenery and variety. As it is almost the most northerly in England and on the extremity

River Lock in Leeds, at the junction with the Aire & Calder Navigation. Bridgewater Place towers above everything else.

of the canal network, it is much less frequently visited than popular routes to Llangollen and Oxford. Although it is a broad canal it has short locks that keep out some of the longer narrowboats. The broad beam reduced the impact of the railways. Two of the canal's wide boats are on show in the National Waterways Museum, Ellesmere Port.

The fastest commercial crossing was in 52 hours, the towpath has been run in 35 hrs 5 mins and it has been cycled in a day.

The canal leaves the Aire & Calder Navigation at River Lock in Leeds, an area where otters are present. Locks and bridges are mostly of dark millstone grit. The bridge arch is often picked out in white and the centre point of the channel is marked, usually offset from the keystone at the middle of the arch, giving a lopsided look.

Overlooking the junction is the Hilton tower block, fine warehouses, Granary Wharf with its Victorian shopping arcade, a 19th century wharf crane and the large Leeds station. The elevated railway hides the cathedral and museum. The indoor market is claimed to be Europe's largest.

The canal follows the River Aire closely all the way to Gargrave, initially running over coal measures. The *ladenses*, British Celtic for a violent river, gave the city its name.

The Leeds to Harrogate railway crosses after Spring Garden Lock. The Leeds Industrial Museum follows at Armley Mills in what was the world's largest woollen mill of 1806 in Georgian stone, used for fulling. It is now has steam locomotives, static engines, water wheels, textiles, a mill cottage and a gas-lit 1920 cinema.

Across fields of rhubarb stand the remains of Kirkstall Abbey, dark and roofless but one of the most complete medieval monastic sites in Britain. It has a good west portal. It was built in 1152–1182 in transitional Norman style by a breakaway group of Cistercians from Fountains Abbey, who rejected Barnoldswick because of its climate. It has the ghost of a 12th century

abbot, who met an unfortunate end. It also has a folk museum in the great gatehouse, with three streets of 18–19th century cottages, workshops, shops with period furnishings, 1760 costumes, toys, domestic items and items from the abbey excavations. Kirkstall Abbey crypt was painted by Turner in 1824.

Newlay Three Rise Locks are another staircase. The minor road to Horsforth crosses the canal and then the river on an attractive cast-iron bridge of 1819, the Micklefield Estate toll bridge, similar to the Scarborough Spa Bridge. It is 3m wide with two 900mm footways and spans 25m with rhomboid openings.

After the A6120 crosses, the Railway warns that the tracks are returning and will cross beyond Calverley. *A Yorkshire Tragedy* was written

Distance
208km from the Aire & Calder Navigation to the Albert Dock
Highlights
Salt's Mill and United Reformed Church, Saltaire
Bingley Five Rise Locks
Belmont Wharf and Springs Branch, Skipton
East Marton double bridge
Pendle Heritage Centre and Weaver's Triangle
Burnley Embankment
Wigan lock flight and replica Wigan Pier
Aintree Racecourse
Albert Dock
Navigation Authority
Canal & River Trust
Canal Society
Leeds & Liverpool Canal Society
www.llcs.org.uk
OS 1:50,000 Sheets
102 Preston & Blackpool
103 Blackburn & Burnley
104 Leeds & Bradford
108 Liverpool
109 Manchester

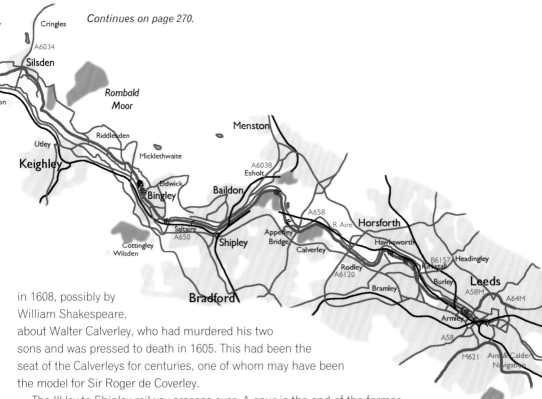

Continues on page 270.

in 1608, possibly by William Shakespeare, about Walter Calverley, who had murdered his two sons and was pressed to death in 1605. This had been the seat of the Calverleys for centuries, one of whom may have been the model for Sir Roger de Coverley.

The Ilkley to Shipley railway crosses over. A spur is the end of the former Bradford Canal. The Noble Comb public house and a high chimney are among the landmarks. In Shipley, the Alhambra had its dance floor replaced with timbers from the short boat *Cedric*. Shipley Wharf has been restored. Canopies still project over the water. This textile and engineering town has 16–17th century mills, a battlemented Salvation Army citadel and Windmill Manor, which was owned by Hilaire Belloc.

Saltaire was a model village, built from 1850 by Sir Titus Salt with 850 good terraced houses for the workers in his mohair and alpaca mill, now a World Heritage Site. The 4ha site has a school, institute and almshouses modelled on Italian villas but no public house or pawnshop. Salt's Mill of 1853, the Palace of Industry, is a six-storey, 22m high building, 166m x 15m, and the upper floor was thought to be the largest room in Europe. The Italian-style buildings include a chimney disguised as a campanile, modelled on Santa Maria Glorioso in Venice. An adjacent single-storey, 100m weaving shed had 1,200 looms and there was a 64m x 34m combing shed. These days there is the largest display of David Hockney works. An adjacent United Reformed church of 1859 is a particularly splendid building. The Victoria Hall has a reed organ and harmonium collection.

On the other side of the river, the Shipley Glen Tramway dates from 1895 and is Britain's oldest working cable tramway, operating to the Bracken Hall Countryside Centre. The park is said to have been one of the most beautiful

Mills before Oddy Two Rise Locks.

in the world and was one of the world's first amusement parks until the tow on a toboggan ride broke in 1900 with fatal results. The four lion sculptures that guard Victoria Hall, near the grammar school, are ones that were rejected from London's Trafalgar Square as they were thought to be too small.

At Cottingley in 1917, two young girls claimed they played with fairies by the waterside. They borrowed a camera, asked to be shown how it worked and duly came back with pictures. They did the same again in 1920. Experts could find no evidence of forgery and Sir Arthur Conan Doyle even said he could see movement in these pictures, which baffled the nation. In 1983 one of them admitted it had been done with cutouts on hat pins but the other died still claiming there was no trickery.

Bingley was a medieval wool and market town, chartered in 1212 but named after the Old English man Bynna, Bynnings *leah* being his people's glade. There is an ancient market cross, Georgian market hall and Bronze Age round barrows. The 16th century Holy Trinity church, with its massive spire, has an early Norman fort and fragments of a Saxon cross.

Then comes another remarkable set of structures: the Bingley Five Rise Locks. These are the country's highest and steepest staircase, lifting the canal 18m. At the top the Five Rise Lock Cafe & Store replaces early stables. Ash trees attract grey wagtails and chaffinches.

Houses alongside the B6265, which follows the canal, hide East Riddlesden Hall. This 17th century Jacobean merchant's manor house is in a small walled garden with beeches and a 17th century 37m oak-framed medieval tithe barn, one of the finest barns in the north of England, near a monastic fishpond that supplied Bolton Abbey. Built in the Civil War by passionate Royalists, the house has a host of ghosts, Yorkshire oak furniture and panelling, wonderful plasterwork, embroidery and pewter. The last three Murgatroyds, the owners, ended up in York Debtors' Prison in Charles II's reign.

Keighley is a manufacturing town. The Cliffe Castle Museum is in a Victorian mansion that was owned by worsted manufacturer Henry Butterfield. It features information on the the history of Airedale, natural history and geology, craftsmen's workshops, reception rooms with French furniture and art treasures and extensive gardens with a conservatory and aviary. Keighley was another Old English man's glade, this time Cyhha. It was home of poet Gordon Bottomley and had a circulating library from which the Brontë sisters were encouraged to borrow books.

Parson's Bridge, beyond the White Lion in Kildwick, connects two parts of the cemetery of St Andrew's, also called the Lang Kirk because of its 46m length. The church has a weathervane, sundial, carved choir pews and a 14th century effigy of Sir Robert de Stiverton, the lord of the manor. It also has wrought-iron lions on the gate, which, amusingly, are reputed to get down after dark to go to the canal for a drink, nothing at all to do with those who have been to the White Lion for a drink themselves. Over the river is a Grade I packhorse bridge of 1306, one of the best in Yorkshire, 42m long with a 200m causeway to the south, of ribbed arches of hewn stone. It is still in use although it has been widened on the downstream side. The canal has a narrow, skewed aqueduct over the approach road. The A629 arrives alongside to give the roar of traffic for a while.

Continues on page 272.

t first the traffic here was transhipment wagons s this was initially near the terminus.

Skipton is a stone town, approached through stone-building canyon. Skipton – Scip-tun, ceptone or Sheeptown – is the gateway to ne Yorkshire Moors. It is one of Yorkshire's ldest market towns, with a market most days n the Georgian high street. The town's pride is kipton Castle, dating from 1090, one of the best-reserved and most complete English medieval astles, still roofed. It has a Conduit Court, a 15m ong banqueting hall decorated with seashells nd a dungeon. It was built by the Normans to efend against the Scots, was overcome in the 2th century, was rebuilt in 1311 by the 1st Lord lifford and withstood a three-year Civil War iege. Butcher Clifford led the Lancastrians in ne Wars of the Roses and his son, the Shepherd ord, led the men of Craven at Flodden in 1513.)ther features of Skipton Castle include a large itchen, six 14th century round towers, a 17th entury yew and a moat. Behind it is a battery. he castle and a limestone quarry quay can be eached along the Springs Branch from Belmont Vharf, which has the remains of a crane plus hops in old canal warehouses.

The A65 comes alongside for a while, the order of the Yorkshire Dales National Park, and ne canal's scenery of rolling moorland is at its est for the next 16km.

Gargrave is a grey limestone town with the 6th century St Andrew's church and fragments f Saxon crosses, a Roman villa and a prehistoric amp. The Pennine Way crosses before Gargrave House in an attractively groomed area of town nd countryside. The A65 crosses back below the ock at the Anchor Inn and leaves.

The six Bank Newton Locks are some of the nost picturesque in the country, set in Pennine cenery. They have iron hooks for towrope juides, the top lock with mason's marks visible.

Once the canal straightens out, it enters a vooded cutting and is followed by the Pennine

Way for a kilometre, during which it is crossed by a unique double-arched bridge at East Marton. The lower arch carried a packhorse bridge and a higher arch has been built on top to take the A59. The Cross Keys Inn refers to St Peter's church, which has a battlemented Norman tower, a fragment of an intricately carved Saxon cross and a Norman font.

The first Lancashire industrial town of Barnoldswick is entered past an array of nest boxes and pirates. This small Pennine town would not be the expected place to find Rolls Royce Aerospace but they have two factories here, the first in an old cotton mill by the B6252 crossing. Indeed, in the RB engine series the 'B' refers to Barnoldswick. Silentnight have a bed

Salt's Mill at Saltaire.

factory in an old textile mill. The 1920 Bancroft Mill Engine is marked by its chimney and has steaming open days with tours as usual, every month.

At Salterforth, the Anchor Inn predates the canal, the change in ground level resulting in the bedrooms becoming the bar and the bar becoming the cellars. The latter now have an impressive display of stalagmites and stalactites. This was the former county boundary, where wool traditionally gave way to cotton because Lancashire was wetter than Yorkshire and cotton manufacture requires more moisture than wool.

At Foulridge Wharf, the former Colne & Skipton Railway crossed on a trestle bridge. Limekilns have been restored and Foulridge Tea Rooms are in old stables. This is at the end of the Foulridge Tunnel.

The walking route is complex and not clearly marked at all turns. It involves negotiating the B6251 and a series of housing estate roads, rather less safe than going through the tunnel for many people with smalls boats to portage. Indeed, what signs there are avoid following the most direct route, along the B6251, instead leading down to one of the Canal & River Trust's canal water supply reservoirs, which are located around the tunnel. The reservoir is used for sailing and few will want to portage beside it when 800m on the water across to the clubhouse is more direct.

The seven Barrowford Locks drop the canal from its summit level to a 39km pound.

As the lock flight bends round, across the field by the A682 is the Grade II Pendle Heritage

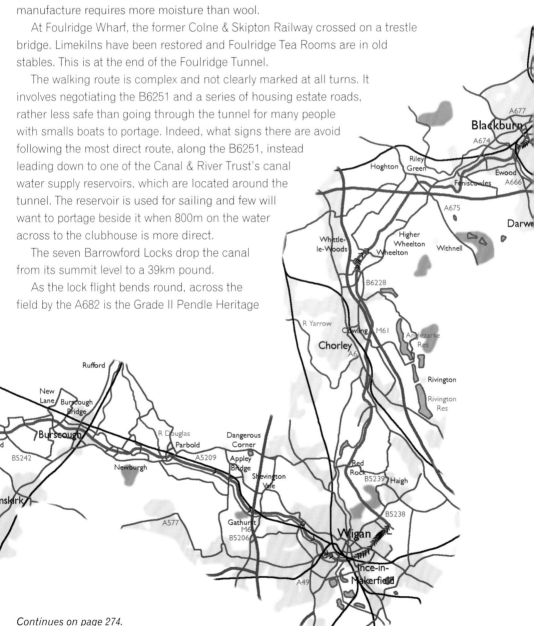

Continues on page 274.

The Bingley Five Rise Locks, Britain's highest lock staircase.

Centre, at one time home to runner Roger Bannister. The centre is 6km from Pendle Hill but on Pendle Water, which the canal is to follow. It features a 15th century cruck barn, a 17–18th century farm labourer's cottage and information about the Pendle witches, ten mostly poverty-ridden females from children to octogenarians who admitted witchcraft and were hung in Lancaster in 1612 after a witch hunt.

The canal crosses Colne Water on the three-arched Swinden Aqueduct, opposite Park Hill. Colne, from the British Celtic *calna*, means noisy river.

The canal winds its way past the terraced houses and weaving sheds of the textile mill town of Nelson, which takes its name from a Lord Nelson Inn in place of its former names of Greater and Little Marsden.

Brierfield is left past a mill with canopies over the water and a church with an unusual clock. The canal is quickly free of the built-up area and meanders along the contours with views to the west across the Pendle valley. Moorhens and mallards paddle about below willows and silver birches. The Colne to Blackburn railway follows the canal, crosses on the approach to Burnley and continues to follow the canal for much of the way to Blackburn. Terraced houses and textile mills return where the coal from Reedley Colliery used to be loaded.

Ahead is another of the wonders of the waterways, the Burnley Embankment, 1.1km long and 18m high. It runs across the valley, wide and straight, perhaps less impressive from above than from below. During the Second World War it had to be drained at times for fear of flooding the town if it was bombed. This is Britain's most outstanding canal embankment. It gives views over the slate roofs of the town, the green copper dome of the town hall, the Comfort Inn and Burnley Football Club's Turf Moor stadium. It also looks away to moors and Coal Clough Wind Farm to the south-east and to 557m Pendle Hill to the north-west, the tautological name of which means hill in Ancient British, Anglo-Saxon and Modern English respectively.

is a safe break in the traffic. The centre is worth visiting if only to see the single tree labelled 'Forest of Pendle', which helps enhance the status of the National Forest further south. In the greater landscape the name remains around the valley of the Sabden Brook to the north, even if the trees don't.

Beyond the tunnel is the Gannow Wharf public house and Rose Grove Wharf, where British Rail had one of their last three steam depots, the engines being filled with water from the canal.

The A646 crosses. The M65 and railway each cross and cross back before the canal winds round Junction 9 and then into a reach where the canal has been moved sideways to allow room for the motorway. A field of caravans precedes the Bridge House Inn at Hapton, with its terraced houses, believed to have been the first street in England to have had electric street lighting, although it is not alone in making the claim.

Eric Morecambe worked at Moorfield Colliery during the Second World War as a Bevan Boy. The A680 crosses by the Albion Ale House in

Gannow Tunnel is much shorter than Foulridge, just 511m long, straight but with no towpath. Steam tugs were used at the turn of the last century and there are mason's marks and symbols on the tunnelmouth. Again, unpowered craft are banned although there is nothing to say so. The walking route is straight up the hill to the A671/M65 Junction 10 roundabout. It is necessary to reach the centre of the roundabout, either down steps and round a tight corner in front of shops or by crossing the road when there

Clayton-le-Moors, the clay in the name being used for brick making. The village was the home of Dr Lovelace's soap factory, which made floating soap. It later had rucksack and outdoor-wear firm Karrimor.

The towpath returns across the canal on Church Kirk changeline bridge, which has had one of its horse ramps replaced with steps. St James' church is reckoned to be the midpoint on the canal. With a 15th century tower and font, Burne-Jones Art Nouveau window, ceiling and balcony on slim pillars and ornate gravestones, it provides a substantial and interesting centrepoint in its setting of terraced houses. The first church was a temporary one set up by Oswald, King of Northumbria, on his way to battle. The current massive tower was used as a watchtower. The village formerly made red dye and printed calico. Sir Robert Peel's family had a cloth works here.

Oswaldtwistle was probably named after Oswald's *twistla*, King Oswald's boundary in the 7th century. It has been an industrial area for over two centuries and its contributions belie its

size. Stanhill resident James Hargreaves invented the spinning jenny in 1764.

Blackburn had Bronze Age inhabitants, the Romans built a fort here and it has been a cotton-weaving centre since the 14th century, much rebuilt since the Second World War. It was a newspaper report on the state of Blackburn's roads that inspired some of John Lennon's most discussed lines.

The Blackburn skyline once had 200 chimneys, one of 95m probably being the highest in the UK. In addition, there are a dozen church spires. The cathedral was the 1826 parish church of St Mary, with medieval and contemporary glass, a magnificent lantern tower restored in 1998, a medieval pax, a John Hayward sculpture of Christ the Worker and a tower topped by a gently tapering cone with a cross on top. Despite the cathedral, Blackburn remains a town. The Museum & Art Gallery in a Grade II Arts & Crafts building has medieval manuscripts, local history,

Belmont Wharf and the Springs Branch.

militaria and early textile machinery, models of the spinning jenny, spinning mule, flying shuttle and spinning wheel. Residents of Blackburn have included George Ellis – the father of the modern brass band – Kathleen Ferrier, Carl Fogarty, Russell Harty and writers Josephine Cox and Alfred Wainwright. Barbara Castle, one of the Kinder Scout open land access trespassers, was MP for 34 years, bringing in legislation on national speed limits, compulsory car seat belts and equal pay for men and women.

The six Blackburn locks bring the 39km Burnley pound to an end. Interest ranges from a toothed segment on a lock gate to a modern statue of a cyclist at Nova Scotia Wharf. There were formerly canal stables. Opposite is the Old Royal Infirmary, the first hospital in east Lancashire, supported by subscription, including that of the canal company. The Atlantic Hotel, Royal Oak and Moorings can provide refreshment in a town that hosts Thwaites brewery.

The Blackburn to Bolton railway crosses before the canal heads out on to Ewood Aqueduct. This crosses the B6447, beneath which the River Darwen is culverted. The large Ewood Park football stadium of Blackburn Rovers is prominent. A further point of refreshment is the Navigation Inn.

To the north of Riley Green, beyond two masts, is Hoghton Tower, a fortified hilltop mansion of 1562–1565, one of the best English Renaissance buildings, owned by the de Hoghton family since the time of William the Conqueror and still owned by them, the second-oldest baronetcy in the country. Shakespeare is said to have worked here for a time as a servant. Better documented is the visit of James I in 1617, when the kilometre-long drive had a red carpet laid and he enjoyed a loin of beef and probably a quantity of wine so much that he drew his sword and knighted the meal.

The unique double-arched bridge at East Marton.

The name has stuck and sirloin steak can be found on the menu of the local Boatyard Inn.

The Top Lock public house is at the start of the attractive Johnson's Hillock Locks, seven locks taking the canal down to join the Walton Summit Branch. Faced with the problem of crossing Preston, the Lancaster Canal stopped on the far side of town and built an 8km tramway to Walton Summit as an interim measure, then continued the canal from Walton Summit to Westhaughton to carry Wigan coal north and bring limestone back. The tramway was abandoned in 1862 and the separate canal section was taken over by the Leeds & Liverpool Canal. Finally, most of the 5km of what had become the Walton Summit Branch was used as the route of the M61 in 1970, leaving just a 300m spur in water. During all this, spa water was found at Whittle Springs, always good for producing travelling customers.

Chorley has had its Flat Iron market since 1498, named after the weights used to hold down its cloth. The town itself was *ceorl leah*, Old English peasants' clearing. It had its market chartered in the 1250s and has become known for Chorley cakes, flat pastry buns filled with dried fruit, a recipe possibly dating back to the Crusades. The most prominent son was Sir Henry Tate, who moved to Liverpool in 1832 to set up his sugar company, also becoming a patron of the arts.

The former coal and textile town of Adlington is the home of Adlington carnival in August. Beyond the White Bear Marina, the largest on the canal, and the Bridge Inn, the Red House Aqueduct takes the canal from Lancashire to Greater Manchester as it crosses the River Douglas. This has already picked up water from several reservoirs including Anglezarke and Rivington.

The canal retains its level as the River Douglas drops away steadily, resulting in ever-more extensive views to the south-west of the canal. The prospect has not gone unnoticed and Haigh Country Park, edged with willows, is set around a pre-Tudor mansion rebuilt in 1830–1849. The park has a golf course, mini zoo, model village, 381mm gauge steam railway, tropical house, formal gardens, nature and geological trails and 1km² of woodlands, largely planted to give work to unemployed cotton workers in the 1860s during the American Civil War.

The 23 locks of the Wigan Flight drop the canal 65m in the better part of 3km, over 5 hours to work a boat through the locks. The Kirklees Hall Inn and Commercial Inn come quickly but the rest of the flight is dry in that sense.

The A573 crosses near the bottom of the flight, followed by the Manchester to Wigan railway and then the West Coast Main Line. Ince-in-Makerfield is named after the Welsh *ynys*, island, British Celtic *macer*, masonry ruins, and Old English *feld*, countryside. A gleaming Girobank building was formerly Wigan power station, the last business to transport coal on the canal. Beyond it is the Leigh Branch. The section from here to Newburgh was originally built as a branch to the Douglas Navigation from the southern route to Liverpool. The River Douglas crosses here and is followed by the canal.

To the Romans, the town was Coccium although Wigan is from the Saxon for rowan trees near a church. It was incorporated by Henry I in 1100 and received its charter in 1246 from Henry III. King Arthur was reputed to have carried out exploits in Wigan. Cromwell pursued the Royalists through the streets in the Civil War. In 1651, the Earl of Derby was defeated in the Battle of Wigan Lane and killed; the mayor is still preceded when out and about by a sword of 1660, given by Charles II for the town's loyalty.

After a playground on the right comes a lock with a dry dock on the left, still in use. Relaunching of small craft is difficult because of the high sides. There is a lower section on the left but it is within a Canal & River Trust area that is kept locked. The next-best option is the far corner of a wharf on the right next to Trencherfield Mill.

Some of the old Weavers' Triangle buildings.

The current mill was built in 1907 to spin cotton imported to Liverpool and brought by canal. It has been converted to apartments but retains the world's largest working steam engine at 1.9MW, which drove an 8.1m, 70t flywheel, powering 84,000 spindles in the mill and used until 1968.

Opie's Museum of Memories in the mill features 40,000 images of 20th century social history. Regardless of its history, it was George Orwell's *Road to Wigan Pier*, written in the 1930s, that made the town's name. Displayed between the A49 bridges, it develops a joke by George Formby Senior in a comparison with the piers of Blackpool and Southport. George was born here, as were Roy Kinnear, Ted Ray, Frank Rendle, Sir Ian McKellan, Angus Fraser, Sir James Anderton and Joe Gormley.

Parbold Hill is 120m and topped by the Grade II Parbold Hall and Ashurst's Beacon, built in

Napoloenic times to be able to warn of French invasion. Reeds build up beside the reach past Priors Wood Hall, crossed near the end by the A5209 as it drops down from Dangerous Corner.

The branch is followed by a cricket field before the Preston to Hunts Cross railway passes over. Burscough was a packet-boat staging post, had stables and was where many of the boatmen lived. Another historical survival until recently was an Easter Pace Egging procession, a local version of a mummers' play.

After a stand of pines, the B5242 crosses Heaton's Bridge, beside which is the Heaton's Bridge Inn. A notice warns of shooting in a beechwood alongside, at the back of which is Scarisbrick Hall. In Victorian Gothic style with a 30m tower, the hall was designed by Pugin for Thomas Eccleston, who drained most of Martin Mere by steam power in 1787 to leave behind a lot of farmland and what has become an important waterbird reserve. Scarisbrick Hall is now a mixed independent school for children up to 16 years.

There used to be a packet boat point by the Blue Elephant Indian restaurant at Pinfold, used by carriages from Southport along what is now the A570, crossing at Scarisbrick Bridge. Once the mere was drained, Wheelwright's Wharf was employed to receive night soil from Liverpool to use as fertiliser, no doubt adding flavour to the leeks grown here. Scarisbrick Cross, made from a single slab of stone, is one of two lines of waymarkers and shrines laid through the marshes to Ormskirk and Burscough Priory.

A road crossing past the Saracen's Head on one side of the canal and a monkey puzzle tree on the other leads to Halsall, where the Grade I St Cuthbert's church of 1290 is the oldest and one of the best in Lancashire. It has a 15th century octagonal tower with spire, original medieval door with an ornate top and a choir vestry that was the grammar school of 1592.

Eanam Wharf was once one of the busiest on the canal.

Views from here are extensive across the flat farmland with its fine soil. Although the canal's inaugural meeting was in Skipton, it was here, in this much easier topography, that digging of the canal began.

Lancashire gives way to Merseyside but the 14th century Scotch Piper Inn may be the oldest in what was Lancashire before the county was

The Orwell has a commanding position in the basin but is not Wigan Pier, despite what the sign says.

Entering the attractive village of Lydiate.

broken up. The A5147 makes a final crossing over Lollies Bridge at the start
of Lydiate, an attractive village of small but well-kept houses with long smart
gardens running down to the canal. More low suspension bridges follow,
including one near the Running Horses public house in Maghull. The Mersey
Motor Boat Club is said to be the oldest of its kind in the country.

The B5422 crosses Red Lion Bridge and then the A59 passes over again. St
Andrew's church of the late 19th century was built in 13th century style to blend
with Unsworth Chapel close by, the oldest church on Merseyside, dating from
1290. Under a beech in the churchyard is the ornate 1936 tomb of Frank Hornby,
who invented Meccano in 1901 and whose company also produced Dinky Toys,
clockwork Hornby Railways and electric Hornby Dublo trains. As such, he did
more for the country's engineering skills base through the 20th century than
anyone else and his contribution to Britain's economy is inestimable.

The M57 crosses. At the end of this straight the canal doubles back at
Aintree, after the River Alt aqueduct, as it is crossed by a very low swing
bridge. Barbed wire ahead indicates the start of the Aintree horse race course,

home of the Grand National. The world's most famous horse race, it has occured annually since 1839. It also has a horse racing simulator, a motor racing circuit that hosted British Grands Prix in the 1950s and, in the centre, the longest nine-hole golf course in the UK.

The name of the settlement is from the Old Norse *eintre*, lone tree. A passenger service by canal to the centre of Liverpool began in 1814. After the A59 crosses Old Roan Bridge and the Preston to Liverpool railway crosses back, there is a stand of poplars. These and willows soften the defences. There are swans and kestrels. The A5036 crosses Netherton Bridge and there are elders, sedges and a thick carpet of azolla in the autumn.

The ornate wrought-iron Litherland Road Bridge with its coat of arms is the first of several interesting road bridges in good decorative order. New housing has been built in the area and the Canal & River Trust have a boat with a lifting bucket on the front like a JCB to excavate shopping trolleys and other debris from the canal. Between the Southport to Liverpool railway bridge and the A5058 bridge there is a coal wharf with the remains of a crane. Red clover adds colour in season. The canal is now running parallel with the docks, 11km of them having been built behind a massive wall along the front between 1824 and 1860, replacing a fashionable seaside resort. By the 1880s Liverpool ships were carrying 40 per cent of the world's trade but the docks still mostly employ the old handling methods and have declined greatly, some now being used by breakers. At Sandhills a large wharf building with an awning over the canal shows that inland water transport was also important.

The Preston to Liverpool railway crosses over and then there is a towrope-damaged post before a wrought-iron bridge, a following ornate bridge with holes in the wall from Second World War enemy fire and another bridge with a rope roller.

The 130m of canal beyond Eldonian Village was bought in the early 1990s by Merseyside Development Corporation, filled in and used as a housing estate site. A further section is dry and walled-in except at the north-west corner. The canal used to continue to Exchange station. It stopped by the A5053, the appropriately named Leeds Street.

The canal now turns right at Vauxhall. Stanley Dock Cut drops through four locks from where there was a connection to the River Mersey from 1846. The last lock is impassable for craft being portaged. Nearby are the former Tate & Lyle sugar factory and a listed parabolic warehouse.

The line passes under the Preston to Liverpool railway although there is no longer a towpath. The Grade II 14-storey Stanley Tobacco Warehouse of 1901 is the world's largest brick building, being restored with accommodation for people and for the pipistrelle bats and falcons that have moved in. The Stanley Dock bascule bridge of 1932 has been restored and precedes the abandoned Collingwood and Salisbury Docks. Turrets top several towers in the outer docks, including a hexagonal clocktower in the centre of the docks, and the warehouses have much character, despite all the smashed windows. In due course the area of restoration and redevelopment must move north along the line of the canal link and these buildings will be made into attractive features. The proposed Liverpool Waters project would take up to 50 years to complete and would include buildings up to 50 storeys high.

The new 690m Liverpool Link accepts boats only six times per day. It passes through the remains of Trafalgar Dock, along the Central Docks Channel, through West Waterloo Dock, over the Kingsway Tunnel, through Princes Half-Tide Dock and Princes Dock (with the Large Objects Collection including steam vehicles and a Blue Streak rocket), under a modern asymmetric footbridge, through the St Nicholas,

Cunard and Museum Tunnels between sections of Pierhead Gardens, over the Queensway Tunnel and past the Beatles Story Pier Head, the new ferry and cruise liner terminal and through the Liver Basin past the Three Graces, which consist of the Royal Liver Building (Britain's first skyscraper with Britain's biggest clock faces), the Cunard Building and the Port of Liverpool Building. A temporary swinging bascule bridge made of 100,000 pieces of Meccano was erected over the canal for a James May TV programme. The new Museum of Liverpool is a lower building but is very obvious from the canal. The route passes Mann Island and through Canning Dock, Stephen Smith's destination from India in *A Pair of Blue Eyes*, with the *Planet*, the former Mersey Bar and Channel light vessel, and the 1900 *Kathleen & May*, Britain's last remaining three-masted topsail schooner. The canal finishes at Salthouse Dock, adjacent to the tidal Albert Dock, location of the Merseyside Maritime Museum with the dock archives, the Beatles Story Albert Dock, the International Slavery

Museum, the Tate Liverpool and various other sports and cultural experiences, including Yellow Duck Tours, which have seen more than one sinking of the amphibious Second World War vehicles, the last time by a tyre in the water. The Pumphouse is now a public house and there are plenty more in the 1846 granite-and-brick Albert Dock buildings by Jesse Hartley. This is the UK's largest group of Grade I buildings, closed in 1972 and now housing shops.

In 2008 Liverpool was the European Capital of Culture and it has been voted the Capital of Pop, the friendliest city and with the best nightlife in the UK. Past writing about the city has included Elizabeth Gaskell's *Mary Barton* and Dickens' *Uncommercial Traveller*. Herman Melville came as a 19-year-old cabin boy, providing the basis for his *Redburn*. Merseyside is the setting for many romantic novels. Kilvert repeated a low opinion of the solvency and honesty of Liverpool merchants in the 1870s. The city has been used for filming *The Virgin of Liverpool, Between the Lines, 51st State, The Hunt for Red October, My Kingdom, In*

The busy and fast-changing skyline in the centre of Liverpool, including the Three Graces, the new Museum of Liverpool, the GWR port facilities, the tug Kerne *and assorted marine equipment.*

the Name of the Father, Letter to Brezhnev, Backbeat and *Priest,* and is the most filmed city outside London. The Beatles, the Searchers, Cilla Black, Gerry & the Pacemakers, the Lightning Seeds and other musicians have had 56 number one music hits, more than any other city. Local artists include Alan Bleasdale, Clive Barker, Beryl Bainbridge, Linda Grant, Adrian Henri, Roger McGough, Jimmy McGovern, Nicholas Monsarrat, Brian Patten, Willy Russell, Ken Dodd and Jimmy Tarbuck. Eleanor Rathbone was the first woman councillor, fighting for better pay and conditions for Liverpool workers, votes for women and, as an MP, the family allowance.

This is the heart of Liverpool, the fishing village of Livpool that received its charter in 1207 from King John and became Britain's largest commercial seaport in Victorian times, handling sugar, spices, tobacco, cotton and slaves. It was the main 19th century gateway for America. The name is from the Old English *lifer*, sludge, but the river is now being cleaned up, following on from the appointment in 1841 of the world's first public health engineer, Dr William Henry Duncan.

The city's 250 monuments and 2,500 listed buildings include the largest collection of Grade II buildings outside London and the city includes Europe's oldest African and Chinese communities.

The new route extends through the historic Albert Dock.

40 Aire & Calder Navigation

The River Aire received an improvement Act in 1699 and it was England's premier navigation by 1704. It is England's leading canal freight route to this day, wide with few locks, able to take barges up to 700t. From the 1860s until 1986 it used Tom Puddings, 40t compartment boats to carry coal, usually towed in trains of 19, to be lifted and tipped into ships. In 1913 there were over a thousand in use. More recently, they were replaced by 210t units, moved in threes by Cawoods Hargreaves, although the exhausting of local coal mines has resulted in the winding-down of the trade. The locks were enlarged from 1884 and push tugs were introduced in the 1960s. Fuel and aggregate transport still runs at about 10,000t per day, so other users should have full competence at boat-handling and keep clear of larger craft.

The navigation begins in Leeds, at the foot of River Lock on the Leeds & Liverpool Canal. The River Aire has just emerged from the Dark Arches beneath the station, a series of long parallel tunnels where the water flows fast with the occasional small fall, the tunnels not being straight and some having complicated exit routes. A roadway and footway run across the insides of the tunnels at high level, parallel to the railway. There are no fish in the river.

Leeds is the UK's second financial city and Yorkshire's second-largest city, having developed in the 14th century on wool and textiles. It was where Michael

Calls Landing is a popular venue in the centre of Leeds.

Marks launched Marks & Spencer in 1884. The Henry Moore Institute is the largest purely sculpture gallery in Europe; Leeds City Art Gallery, Leeds City Museum, the Northern Ballet and Opera North are also resident. The City Varieties Music Hall has been used as a singing venue since 1762. The city has been used for filming *Harry's Game, Jimmy's, The Good Old Days* and *A Touch of Frost*. Leeds has 8km² of parks.

On the south side is Clarence Dock, also known as Tatie Basin from its handling of Jersey potato traffic. Overlooking both the dock and Leeds Lock is the Royal Armouries Museum, Britain's largest post-war museum, which was moved from the Tower of London. It has the world's best collection of armaments, guns and armour, even elephant armour. Demonstrations include jousting, falconry and Wild West gunfights.

A steam derrick crane draws attention to Thwaite Mills Industrial Museum. There was a fulling mill here between the navigation and river. It had four wheels from 1641 and was bought by the Aire & Calder Navigation Company to safeguard their water supplies. The current seed, flint and chalk crushing mill of 1823, one of the last remaining water powered mills, was also used for preparing putty, pottery, bread and medicines at various times and had china clay delivered direct from Fowey.

Rising land to the south hides Rothwell and the former mining village of John O'Gaunts although spoil tips can only be landscaped so far. To the north is Temple Newsam Country Park, one of the largest floral parks and working rare-breeds farms in Europe, 5km² having been laid out in the 1760s by Capability Brown around a Tudor/Jacobean red-brick mansion of c1500. Described as the Hampton Court of the north, it was the birthplace of Lord Darnley and owned by the Knights Templar in the 12–14th centuries. Lakes and country park have been established on opencast workings.

Distance
55km from River Lock to the River Ouse
Highlights
Royal Armouries Museum at Clarence Dock
Thwaite Mills Industrial Museum
Yorkshire Waterways Museum, Goole
Navigation Authority
Canal & River Trust
OS 1:50,000 Sheets
104 Leeds & Bradford
105 York & Selby
(106 Market Weighton)
111 Sheffield & Doncaster
112 Scunthorpe & Gainsborough

The Royal Armouries Museum overlooks Leeds Lock.

Moorings above Goole.

Powerlines cross before the confluence with the River Calder, bringing in the Wakefield Section. The cut becomes the boundary between Leeds and Wakefield. The river loops away towards Castleford and Allinson's flourmills with the world's largest capacity for stone-ground flour, grinding by water power at a weir.

On Ermine Street, from Lincoln to Tadcaster, the Roman fort of Lagentium guarded the Aire crossing point, the most difficult river crossing on the Great North Road. It was the site of a battle between warring factions in 947. There are fort and bathhouse remains and a museum has archaeological finds. There was famous glassware and pottery in the 18th century but Castleford's claim to fame these days is as the birthplace of Henry Moore.

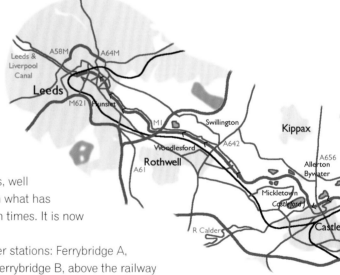

Beyond a bowstring railway bridge and powerlines are *ings*, Old Norse for meadows. Castleford Ings, on the south side, face Newton Ings and Fairburn Ings, well endowed with lakes. The latter is on what has been an industrial site since Roman times. It is now a wetlands reserve for the RSPB.

Ferrybridge is known for its power stations: Ferrybridge A, below the railway since 1927, and Ferrybridge B, above the railway from 1953, have both closed. Today there is just the 2GW Ferrybridge C from 1967, the largest in Yorkshire, burning 1,000t of coal per hour.

After all the former mine sites and power stations, the darkened stone church tower near the river in Brotherton comes unexpectedly, as does Brotherton Bridge, which carries the York to Rotherham railway. Built in 1840 as an impressive tubular bridge on massive abutments, it was replaced in 1901–1903 by towering Whipple Murphy trusses. At one time it formed part of the only line between Scotland and England. The queen and three of her children were said to have been placed here for safety while Edward II was fighting the Scots.

Located so as to be visible through a bridge arch is the traffic light for Ferrybridge Lock, a flood lock normally open beyond the Golden Lion as the navigation leaves the river for the last time and takes to the Knottingley & Goole Canal. Sir Richard at the Lee was said to have crossed the Aire here with Robin Hood's party.

Knottingley was named after the Dane Cnotta. The canal has a gentrified urban feel to it at first, unique for this canal. Near Mill Bridge there is a branch to a weir on the River Aire. The Steam Packet Inn recalls passenger transport on the canal.

Arrival at Pollington is between the imposing hall on the north side and a diminutive but conspicuous church spire to the south. In 1992 Pollington Lock was the first British Waterways lock to have gates made of West African opepe instead of oak and greenheart. Enlarged several times, it has three chambers and is one of the largest locks on the inland waterways.

The canal has been converging on the River Went but never actually meets it. Instead, it is met by the New Junction Canal. Built in 1905, it was the last new canal in England, until recent years, and runs dead-straight for 9km.

Formerly an industrial eyesore, Sugar Mill Ponds have become a local nature reserve. Rawcliffe Bridge is served by the Black Horse, near which there is a mooring area. The latter is across the canal from a mill that emits a pungent odour when the wind is from the north. The Wakefield to Goole railway arrives from behind the mill, to run beside the canal. A pheasant may fly low across from one bank to the other.

One of the first premises on the south side of Goole Docks is the Yorkshire Waterways Museum. It gives an account of the canal's history and offers a cafe. The docks were created in 1828 for the canal, to export coal and textiles, and are Britain's premier inland port. Although they are the furthest inland port they trade with the Baltic and other parts of Europe. A Tom Pudding tippler, virtually unchanged since 1865, was in operation until 1986, one of four used on land and accompanied by another floating one.

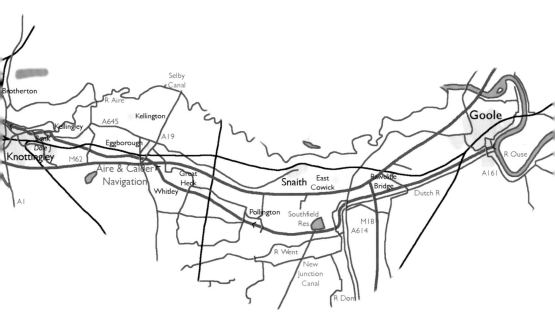

41 Aire & Calder Navigation, Wakefield Section

The Wakefield Section of the Aire & Calder Navigation runs north-east across West Yorkshire from Wakefield to Castleford. It links the Calder & Hebble Navigation with the main line of the Aire & Calder Navigation, part of the trans-Pennine route to the Humber. The Calder & Hebble Navigation ends at Fall Ing Lock at Belle Vue. The Wakefield Section uses the River Calder at first. On the opposite side is Wakefield Old Lock at the end of the short Old Wharf branch.

High ground on the right, topped by the hall at Heath, pulls back. A railway bridge crosses. The Southern Washlands Nature Reserve is located in a former opencast colliery. The navigation carries on as the river bends right.

Broadreach Flood Lock protects the navigation as the river begins several kilometres of meanders. The navigation then runs straight for over 2km.

The hamlet of Stanley Ferry is the major point of interest on this canal. The Stanley Ferry inn, positioned next to a spur that acts as a marina, is popular.

The jewel in the crown, however, is the arched aqueduct over the River Calder. The original design of six arches was rejected because of spate flow volumes in the river. Instead, George Leather (with assistance from Thomas Telford) came up with this design, one of the

The Calder aqueduct: the largest cast-iron aqueduct in the world and one of the most ornate.

Looking down the the River Calder from Fall Ing Lock.

Distance
12km from Wakefield to Castleford
Highlights
River Calder aqueduct, the largest cast-iron aqueduct in the world (better viewed from the newer concrete aqueduct alongside)
Navigation Authority
Canal & River Trust
OS 1:50,000 Sheets
104 Leeds & Bradford
105 York & Selby
110 Sheffield & Huddersfield
(111 Sheffield & Doncaster)

most interesting aqueducts in Britain. Each side has a cast-iron arch in seven segments, from which the iron trough is hung by 35 wrought-iron rods. Built in 1836–1839, it weighs 1,700t, is 50m long, holds 940t of water and is the largest cast-iron aqueduct in the world. Its full glory is not seen by passing over it, however. It is better viewed from the 2,300t concrete aqueduct built alongside in 1981, from where it is seen to have a dramatic Renaissance style with closely spaced fluted columns along each side and pediments in each of the four corners. The new aqueduct is wider, reducing the potential for collision damage. The Canal & River Trust have a large maintenance yard adjacent. To their credit, they floodlight the aqueduct each night, such is its splendour.

Altofts is an outlier of Normanton, a settlement based on the canal trade. Down the hill from the village is King's Road Lock. Further down on the same side are the remains of Foxholes Side Lock, which served Foxholes Basin, now filled in.

Below the farm at Penbank, which is marked with a sign like that of a public house, the Fairies Hill Lock Cut still exists on the right, formerly taking in Altofts Lock, Altofts Basin and Fairies Hill Lock. These have been replaced by the 4.1m deep Woodnook Lock. This enables the navigation to rejoin the River Calder above the railway viaduct rather than below it.

The Fairies Hill Lock Cut and a colliery basin emerge together on the right, midway between the railway viaduct and another around the following bend. They are followed soon after by a further bridge for a dismantled railway, such was the proliferation of lines in this area. Castleford station now uses a single platform.

The river takes a relatively straight line in the vicinity of Castleford Mere, cutting between various oxbow lakes in former meanders. A tank farm is passed on the right before arriving at a water crossroads with the River Aire at Castleford Junction.

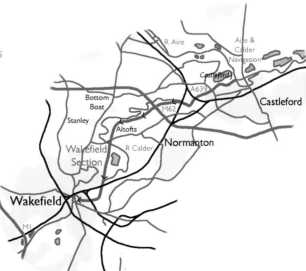

42 Leeds & Liverpool Canal: Rufford Branch

From 1740 until late in the 18th century boats wishing to travel from Wigan to Preston did so via the River Douglas. The Rufford Branch of the Leeds & Liverpool Canal, also known as the Lower Douglas Navigation, provided an alternative route for much of the distance. It was opened in 1781, long before the main line of the Leeds & Liverpool Canal was completed in 1816.

Running north, it drops quickly at first from the main line at Burscough Bridge to cross Burscough, Mawdesley and Croston Mosses. This is flat but fertile farmland, which forms the Lancashire coastal plain, a completely rural route.

Turning off the main line, the Rufford Branch passes under an imposing arched bridge of 1816 and immediately enters the most interesting area on the branch, the canal settlement around the basin at Lathom, a conservation area. On the left is a dry dock with all its boat support saddles in place. The dock is filled by allowing water in from the basin and emptied by permitting it to drain out further down the lock flight, a very neat and simple system.

Beside Lathom Bottom Lock is the Ship Inn. It has a rather higher class of customer than in earlier days, when it went under the name of the Blood Tub.

The bywash at each lock is unusual in that it is exposed and runs past the lock

The lock flight at Lathom. The end houses are cut away to assist towropes.

in large section culvert at high level, discharging down to the lower level in the vicinity of the bottom gate. Paddle gear for filling locks is also unusual, taking the appearance of a couple of miniature balance beams above the top gates, which have to be lifted to open them, a process that is complicated by their having to be padlocked down to prevent non-authorised interference. Some of the lower locks, instead, have a screw arrangement, using a very coarse pitch of threaded rod.

After the penultimate lock at Rufford, the canal passes a small Italianate Victorian church with many monuments to the Hesketh family, who were local landowners. One sculpture, dating from the late 1450s, memorialises no less than 11 offspring.

On the left is Rufford Old Hall, a timbered house, the finest 15th century building in Lancashire, with Jacobean extensions; ornate hammer-beam roof; an intricately carved, immense oak screen; fine collections of 16th century arms and armour and 17th century oak furniture; as well as a folk museum, 6ha garden, shop and tea room. The house was once owned by the Hesketh family. Shakespeare is thought to have performed here. There used to be a swing bridge opposite the house. Although it has been removed, the route is still used by the Old Grey Lady, a spectre crossing from the Old Hall.

> **Distance**
> 12km from the main line to the River Douglas
> **Highlights**
> Lathom conservation area
> Rufford Old Hall, the finest of its kind in Lancashire
> **Navigation Authority**
> Canal & River Trust
> **OS 1:50,000 Sheets**
> 102 Preston & Blackpool
> 108 Liverpool

The Grade II Rufford New Hall dates merely from 1760, the Heskeths moving here from the Old Hall. It still retains its icehouse in the garden.

The final lock has diamond gates at the bottom to act as tidal doors. These only open on to the tidal River Asland or Douglas at high water to give access to the River Ribble and Preston or the Irish Sea. It is significant that the banks are high, steep and muddy and that the tidal river flows very fast. Timing is important for ongoing trips and advice is given locally.

German's Lock on Burscough Moss with extra balance beams.

Tarleton

Sollom

Croston

A59

A581

R Douglas

Rufford

Rufford Branch

Burscough Bridge

Leeds & Liverpool Canal

Burscough

43 Lancaster Canal

The Lancaster Canal ran southwards through Cumbria and Lancashire from Kendal to Preston, a broad-beam contour canal designed by Rennie and constructed between 1797 and 1819. It had been intended that it should join the Leeds & Liverpool Canal at Wigan but the final link across the River Ribble was never built and the canal remained isolated, except for a branch in the centre to the sea. Promoted by Lancaster merchants, it carried coal, lime, slate, timber and food. It also carried passengers. Despite the hilly terrain, Rennie managed to produce a 92km run with only a flight of eight locks at Tewitfield and the country's longest level pound below, 68km. This allowed express packets to travel from Preston to Kendal with frequent changes of horse, averaging 16km/h including changing horses, in the astonishingly quick time of seven hours in 1833.

The service ceased in 1846 because of rail competition. The last coal barges ceased their trade 101 years later and the canal was abandoned above Tewitfield in 1955. Culverting of several points above this, mostly by the M6, means that reopening of the canal will be more difficult than would previously have been the case. There is little midweek activity on the canal.

Despite being on the edge of the Lake District, the scenery is not as hilly as might be expected except at Farleton Fell, an extensively quarried limestone crag

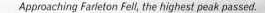

Approaching Farleton Fell, the highest peak passed.

rearing up some 400m above the canal, peaking only 800m away. Farleton was an important canal village, its stable and packet house now disused. The former coaching inn of 1630 by Duke's Bridge has become a private house.

The canal has been all at one level so far. Now the eight stone-chambered locks drop the canal 23m to its other long level pound, which reaches all the way to Preston. Was Rennie just lucky or was it skilful engineering that enabled him to group all the locks in 1km? Perhaps a solution to the level problem at the motorway crossing is to rebuild the top lock to the west of the motorway and deepen the cut for several hundred metres.

The final culvert comes immediately below the bottom lock. The A6070 crosses on a high embankment next to the Longlands Hotel. The towpath is diverted under the A6070 bridge across the M6, to emerge on the far side in Tewitfield Marina. Here there are narrowboats moored right up to the final obstruction, picnic tables, a children's playground and full Canal & River Trust facilities.

The canal moves away under Tewitfield Turnpike Bridge towards Borwick. Borwick Hall was an Elizabethan manor house built around a 15th century peel tower, the gateway including a stone dating it at 1650. Charles II stayed here in 1651. These days it is a Lancashire Education Authority youth club training centre.

Warton Crag towers high above with Warton at its base, including the 15th century tower of St Oswald's church, which has stars and stripes on the coat of arms of the family ancestors

Distance
81km from Stainton to Preston
Highlights
Lune Aqueduct, as painted by JMW Turner
Carnforth Railway Museum with the *Flying Scotsman*
Navigation Authority
Canal & River Trust
Canal Societies
Lancaster Canal Trust
www.lancastercanaltrust.org.uk
The Lancaster Canal Restoration Partnership
www.thenorthernreaches.co.uk
Ribble Link Trust Ltd
www.ribblelink.co.uk
OS 1:50,000 Sheets
97 Kendal & Morecambe
102 Preston & Blackpool

of George Washington, interestingly preceding their use in the American flag.

Signposts on the canal point not only along the water channels but also to town centres and features of interest, as does the one by the playground in Carnforth. Carnforth is very much

Rennie's masterpiece, the magnificent Lune Aqueduct, carries the canal over the River Lune in Lancaster.

a railway centre, the point where the Furness and Leeds lines depart from the West Coast Main Line. Steamtown was one of the last depots servicing steam engines during the change to diesels in 1968 and is a railway museum with over 30 steam locomotives including one of the most famous, the *Flying Scotsman*, signal box, turntable and narrow- and standard-gauge lines. In 1945 the station was used as the location for filming *Brief Encounter*, adapted from Noel Coward's *Still Life*.

The A6 and then the railway crowd in on the canal. There are fine views past a caravan site and out over the wide expanse of Morecambe Bay to the southern fells of the Lake District. Inevitably, this section of canal is exposed to westerly winds.

The canal meanders past wooded hillsides and parkland at Bolton-le-Sands, the Royal Hotel and the Packet Boat Hotel by a former wharf.

At Bolton Town End, a school complete with bell stands beside the canal. The A6 crosses and goes on its way. Hatlex swing bridge is usually left blocking the canal. The high-water mark at Hest Bank is only 200m from the canal. It is hidden from the cut by houses at its closest point, a nature reserve. The bay is one of the top five wildfowl feeding areas in Europe with up to 50,000 waders visible, some of which find their way on to the canal. A stone pier, previously used for transhipping freight between barges and coasters, became disused after 1831 when Glasson Dock was established. A footpath across the Kent estuary to Kents Bank, probably much older, still runs out from here. The Hest Bank Hotel is an old coaching inn standing near the canal.

The A6 recrosses before the canal turns sharply past an overflow outlet to the River Lune and crosses one of the finest aqueducts in

Tall brick mill buildings crowd the canal in central Lancaster.

Britain. Built in sandstone from 1794 to 1797, it has excellent proportions and was Rennie's first aqueduct incorporating hidden inverted arches to spread the stresses and the first to have internal iron bars reinforcing the stonework. The foundations were made with pozzolana imported from Italy. The Grade I 202m Lune Aqueduct stands 19m above the river on five 21m semi-circular arches that rise 18m from the riverbed.

Curved wing walls and Gothic ends on rustic pillars complete the canal's major engineering work, a structure that Turner chose to paint. Bulk Road Aqueduct, by way of contrast, crosses the A683 and was only built in 1961.

Rising above the golf course at its end is the Ashton Memorial of 1907–1909, built in neoclassical style by Lord Ashton in the 15ha Williamson Park of 1896 in memory of his wife. The Edwardian era is shown on screen, the Old Palm House has tropical butterflies and Dukes Playhouse has films and theatre, the Promenades being among the most spectacular open-air theatres in Britain. Uphill from the large tidal weir on the Lune, the canal enters the fully built-up part of the city.

Lancaster is the red rose city, centre for the Lancastrians in the Wars of the Roses. It takes its name from *lune*, Celtic for healthy, and *ceaster*, Old English for fort. The market square is where Charles II was proclaimed king in 1651. In the 19th century local palaeontologist Sir Richard Owen planned London's new Natural History Museum and invented the word 'dinosaur' but claimed Darwin's natural selection theories did not fully explain evolution.

The Romans built a bath house in Lancaster. The 15th century Perpendicular Benedictine Priory church of St Mary includes a Saxon wall and doorway on the site of a Roman fort and contains 13th century carved choirstalls, which are some of the earliest and finest in England, in addition to fine needlework and Abyssinian Coptic crosses. A Saxon church of about 600 was

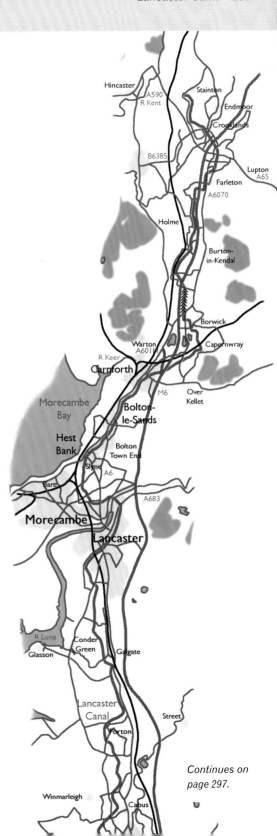

Continues on page 297.

replaced with a 1380–1430 model, the memorial chapel of which has the colours of the King's Own Royal (Lancashire) Regiment.

The Norman castle is now a prison, past residents including the ten Demdyke witches of 1612 and Quaker founder George Fox, its implements including a clamp last used for branding criminals in 1811 and a cat o' nine tails last used in 1915. The Turret or John of Gaunt's Chair has a view to the Isle of Man and was used to signal the approach of the Armada. The massive gatehouse is 15th century but most of the castle was restored in the 18th and 19th centuries. The county court uses a 19th century room. Its two-storey keep was heightened from its original form and there remains part of a bailey curtain with a round and two square towers.

Less dramatic buildings include the Baroque Music Room of the 1730s with ornate plasterwork, the Grade I Judges' Lodgings town

house with Gillow furniture and a museum of childhood. The Georgian former town hall of 1783 contains the museum of the King's Own Royal Regiment (Lancaster). The city museum and the Cottage Museum are in an artisan's house of about 1820. St Peter's Roman Catholic church of 1859, in Geometrical style, was one of Paley's finest, with ten bells, notable architecture and stained glass, as well as one of the finest organs of its type. It became a cathedral in 1924.

By the infirmary is the Canal & River Trust depot with its hand-operated crane. The West Coast Main Line crosses for the last time but is to follow the canal for much of the rest of its journey to Preston. Also crossing is Aldcliffe Road footbridge, made in 1954 from a shortened ship's gangplank, with a postbox in the road side of the bridge.

The canal moves south from Lancaster past a neat fence dividing it from an adjacent road. It passes an old castellated brick boathouse,

Old canalside buildings refurbished at the White Cross.

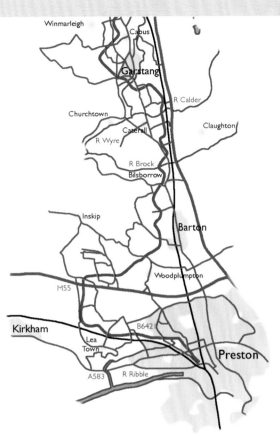

smothered in ivy, as it enters the 3km long Burrow Heights cutting. In the centre of the cutting is Broken Back Bridge. It takes its name from the unusual lines of its stone courses. Hidden is Burrow Beck, which passes beneath in a syphon. During construction of the cutting, a Roman find of two lions, four heads and a stone statue of Ceres was unearthed, now in the Lancaster City Museum.

From the bridge in Galgate, the canal crosses the Conder Aqueduct and winds round a cricket pitch to meet the A6 and the railway. The latter frames the far side of the village with high embankment, bridges and a viaduct. Beyond the viaduct is the tower of the church of St John, by which is the oldest silk-spinning mill in England, built in 1792. The stone village comes to an end with Galgate Basin Wharf picnic table area, moorings and basin excavated in 1972–1973. Canalside moorings are extensive with boats moored at an angle on the east side as well as parallel along the west bank.

At Lodge Hill, the Glasson Dock Branch of 1825 leaves to run down the Conder valley to Glasson and the Lune estuary, until recently the canal's one contact with the outside world. At the junction are a narrow stone bridge and a lock keeper's cottage, frequent locks on the branch contrasting with the absence of them on the main line.

Opposite Junction 33 of the M6 is the fine Italianate villa of Ellel Grange, built 1857–1879 on top of a ridge. The adjacent church of St Mary is charming. Local bridges include an ornamental one with balustraded parapets across the canal and a double bridge with a wall down the middle to separate farm and estate roads.

An aqueduct carries the canal over the River Cocker, which flows down through Bay Horse where a Euston to Glasgow train ran into a local train in 1848 when two different companies were running their trains over the same lines.

From Potters Brook can be seen the distinctive mushroom-shaped tower of Lancaster services on the M6. The canal is crossed by the Lancashire Cycle Way at Forton. Forton's stirring contribution to the war effort during the Second World War was the removal of the parapets of Stony Lane bridge so that enemy troops might be seen crossing it. Views are frequently extensive from the canal. From here, the expanse of the Fylde, the coastal plain that is Lancashire's market garden, can be seen. Inland, the fells are steadily rising again around the Forest of Bowland.

Tank traps beside the canal at Ford Green show that the canal's defensive role was taken very seriously. Brick bridge abutments by the canal just before Garstang are those of the former Garstang & Knott End Railway, alternatively known as the Pilling Pig.

The canal passes under the A6 once again at Cathouse Bridge, Garstang formerly having been a staging post on the A6. Among other bridges, it passes under a pipe bridge of 1927, which carries water in a sweeping arch on its way from Barnacre Reservoir to Blackpool, the former of which can be located by a trio of aerials. Thursday is Garstang's busy market day, the market having received its charter from Edward II in 1310. The town hall with its diminutive belltower, opposite the cobbled market place, dates from 1680. The church of St Thomas is an 18th century structure. The Owd Tithe Barn, in stone and timber beside the canal, was restored and reopened in 1973 as a canal and agricultural museum and restaurant. It was in Garstang that an ice-breaking barge sank in 1945.

The canal leaves Garstang over the Wyre Aqueduct, a single elliptical 16m stone arch passing 10m above the River Wyre. It passes round Bonds to turn just short of Greenhalgh Castle. This was built in 1490 by the Earl of Derby and was a Royalist stronghold until it was destroyed by the Roundheads in the Civil War. Now it is largely ruined.

Powerlines pass backwards and forwards over the canal for some distance. More traditional power is seen at a coalyard where the railway and M6 come back alongside the canal for a final 3km.

Catterall Basin was formerly the site of a papermill. These days a wooden bench seat suspended on ropes as a garden swing beside the canal suggests a more leisurely pace of life. The River Calder is passed in a siphon under the Calder Aqueduct.

Beyond an aerial in Catterall is the tower of St Helen's in Churchtown, a magnificent parish church dating in parts from about 1300 and known as the Cathedral of the Fylde. The Grade I Claughton Hall was originally a church neighbour. All except one wing was dismantled and moved to its present hilltop site.

First comes the Brock Aqueduct, where a levelling error resulted in a tunnel having to be built under the canal for the River Brock. Just down the river valley stands the Myerscough College on the outskirts of Bilsborrow.

The Lancashire Cycle Way crosses just before Hollowforth Aqueduct, which clears the Barton Brook. It is followed by a swing bridge that is closed by means of a chain lying on the bed of the canal.

The canal crosses the approach line of the main runway at Warton Aerodrome, where the Typhoon or Eurofighter was built, tested in the mountains of the Lake District to simulate conditions in Afghanistan.

Easing its way into Preston, the canal passes Haslam Park and crosses the Savick Aqueduct. Proactive canal enthusiasts have canalised the Savick Brook to provide a canal link with the rest of the canal system via the River Ribble, River Douglas and the Rufford Branch of the Leeds & Liverpool Canal.

The city largely ignores its canal and does not show its best face. A slender church spire near Moor Park is striking. Preston was the second-oldest borough in England with a Member of Parliament since the 13th century and a Merchants Guild that has met every 20 years since 1542. It was a market town that developed in the Industrial Revolution, not least because Arkwright was born here in 1732 and his spinning frame of 1768 made the town a cotton-spinning centre for 150 years.

An ornate bridge across the canal at Ellel Grange.

44 Stainforth & Keadby Canal

The Stainforth & Keadby Canal was built between 1793 and 1802 to extend the River Dun Navigation from near Stainforth to the River Trent. Both form parts of the Sheffield & South Yorkshire Navigation. Effectively, it was a ship canal for 200t craft and was to prove a financial success. Being cut across fenland, it had mostly long straight reaches and only needed two locks, the second of which was to deal with the tides on the Trent. Construction of the New Junction Canal has now removed most of its traffic.

It is a waterway heavily used by anglers, having numbered pegs over much of its length. Most bridges open and are low.

At first the canal follows the tidal River Don, on the other side of which is Fishlake with an armless windmill and the fine medieval church of St Cuthbert. The church is noted for its intricately carved late-Norman doorway with animals and foliage and for a belltower with a notice banning the ringers from wearing hats or spurs.

Beyond the M18, the Doncaster to Hull railway crosses as the canal enters Thorne, the only town along its line. The Canal Tavern is tucked in next to the A614 as it crosses and the Doncaster to Grimsby railway line crosses and follows the canal, most of the way keeping close to the bank of the canal. Fenland is opening up, flat but fertile, and plants include bog rosemary.

Thorne Lock is one of only two locks on the canal.

To the north and south respectively are Thorne Waste and Hatfield Chase, former marshes, the largest and most diverse lowland peat area in northern Europe, now mostly cut commercially but still a habitat for rare plants and birds.

The Old River Don is crossed, now no more than a drainage ditch, for which it would be hard to guess the original line. The diversion to the current line was undertaken by Vermuyden in 1625, resulting in the name Dutch River.

The canal avoids Crowle with its Georgian houses. Its church has a large chancel, a clerestory and a 2.1m Saxon carved stone, believed to represent King Oswald and his son, Oswry. The Regal Motor Museum is housed in a former cinema. It has cars and motorcycles of 1902–1930 and a Flying Flea aircraft of 1936.

Approaching Keadby, there is a minor swing bridge and then the railway makes an oblique crossing on the Vazon sliding bridge designed in 1926. The heavy girder structure rolls sideways, taking the track with it. As it closes, the track moves back to its position. Accompanied by assorted rumblings and clicks, the track is pulled down some 100mm into place and rods slide into position to lock everything together. This is a marvellous piece of machinery to watch in operation, something that does not happen frequently.

Keadby Lock is a two-way structure because the River Trent, on to which it opens, is tidal. The river has streams that may exceed 11km/h and has an eagre on equinoctial spring tides. Although the lock length is limited, longer boats can pass through when the Trent and the canal are at the same level so that all gates can be opened for clear passage through.

Also seen from the lock is the King George V Bridge over the Trent above Althorpe. Built in 1916, it carries the railway and the A18. It was the Great Central Railway's largest bridge project. With five spans, it includes an unusual 50m rolling lift bascule span, the heaviest bascule span in Europe, fixed down in 1960.

Distance
20km from the River Dun Navigation to the River Trent

Highlights
St Cuthbert's church, Fishlake
Rigle Motor Museum, Crowle
Vazon sliding bridge near Keadby

Navigation Authority
Canal & River Trust

OS 1:50,000 Sheets
111 Sheffield & Doncaster
112 Scunthorpe & Gainsborough

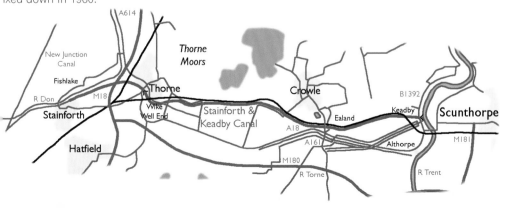

45 Chesterfield Canal

Despite being one of the early canals, the Chesterfield Canal or Cuckoo Dyke included some notable engineering. In particular, it needed 49 locks in the top 32km in order to get over a ridge of high land from the Rother valley. It also needed the Norwood Tunnel, still the sixth-longest to have been built in Britain. The canal was designed and constructed by Brindley although he died before it was opened in 1777. It was two years late, following problems with the tunnel, which was 50 per cent over budget because of fraud. Traffic peaked at 200,000t in 1848, particularly stone, grain, lime, timber and ale, but decline followed quickly after it was purchased by the Manchester & Lincoln Union Railway. The section above Worksop was unusable by 1896 and the tunnel suffered a collapse in 1908 as a result of mining subsidence. All commercial traffic ceased in 1962 although the earlier closure of Walkeringham brickworks had been a major blow to commercial use.

The isolated canal runs eastwards from Chesterfield to the River Trent. It is a narrow winding canal, mostly

The pumping station by the Ryton aqueduct.

quiet and rural and often passing through attractive wooded scenery. Much has been restored.

The canal leaves the River Rother but follows its right bank for several kilometres. Tapton Mill Floodgate is at a small brick bridge. The view back to Chesterfield (named from the Old English or a Roman fort in open country) is crowned by the twisted and bent spire of St Mary's church, which was hidden at the start.

A recent tunnel carries the A619 across just before Ford Lane Lock. The lock itself, together with its lock cottage, now a canal visitor centre, has the feeling of being in the middle of a busy roundabout although traffic does not go all the way round. Artwork is displayed under the first bridge. The Cuckoo Way long-distance footpath follows the canal and the Trans Pennine Trail also uses the towpath to Staveley.

Mosaic panels are set into the towpath by Wheeldon Mill Lock. Between the canal and the river there was a railway at one time. Blue Bank Lock faces on to a reach that is rather open. Between New Whittington and Brimington was the Dixon Mine. Dixon Lock has been restored with a tiny humped footbridge, which is typical of this canal.

Restoration of the eastern end of the canal has been completed up as far as the eastern portal at Kiveton Park, now bricked-up. The canal is followed closely by the Sheffield to Cleethorpes railway. After a disastrous fire in 1834, 32,000t of stone was exported from here for rebuilding the Houses of Parliament.

A winding hole appears alongside Old Spring Wood before the descent begins in earnest: 22 locks in 1.6km.

The canal eases into Worksop, which developed from the Old English Weorc's valley to serve the North Nottinghamshire Coalfield. Surrounding woodland is the remains of Sherwood Forest.

Just downstream of Worksop Lock is the

Distance
81km from Chesterfield to the River Trent

Highlights
Hollgwood Hub canal visitor centre
Turnerwood lock flight
Cuckoo Wharf and Pickford's Warehouse
Lady's Bridge (also known as Old Man's Bridge)
West Stockwith Basin

Navigation Authority
Canal & River Trust

Canal Trust
Chesterfield Canal Trust
www.chesterfield-canal-trust.org.uk

OS 1:50,000 Sheets
(111 Sheffield & Doncaster)
112 Scunthorpe & Gainsborough
119 Buxton & Matlock
120 Mansfield & Worksop

restored Cuckoo Wharf with a three-storey Pickford warehouse built over the canal, now featuring a canalside bar with tables round a traditional canal crane.

There are some interesting old buildings on the right, behind the Lock Tavern. These include Worksop Museum and the priory, which has a 12th century restored church and a 14th century gatehouse containing a skull in which an arrowhead is embedded.

The views from the canal now become extensive over parkland. Owned by the Duke of

Section past a wooded bank near New Whittington.

Newcastle, who was one of the contributors to the name of the Dukeries for the area, Clumber Park's 10km² of country park reach nearly to the canal. Close on the north bank is the Grade II Osberton Hall and its associated buildings with some graceful architecture and farmland in immaculate order.

From the Chequers Inn the canal moves back to be followed by the A1 for a kilometre. This is perhaps not as noisy as might be expected because of the screen of trees.

A series of three aqueducts includes one over the River Idle, the valley of which is followed over the rest of the canal's course. Old wharf buildings are accompanied by modern buildings in complementary design on the East Retford side of the river by Retford Lock, the red ford named after the clay in the River Idle. The Wharf restaurant follows just before the canal passes under the old Great North Road bridge, now reduced to the status of a minor town street. The canal passes King's Park. Notable town buildings include the flamboyant town hall of 1868 and a cruciform church with a peal of ten bells and

protection in the form of an 11kg cannon brought back from Turkey.

Whitsunday Pie Lock is the first of the wide locks. Retford helped finance the canal to Retford on condition that it was built wide enough to bring Trent keels up to the town. The lock is said to take its name from a huge pie baked for the navvies by the wife of a local farmer to celebrate completion of the lock, a fact recorded by the serving of a pie to boat club members each year. However, there are signs that the name is rather older, relating to the time of the tenancy change of a pightle of land.

Bonemill Bridge takes its name from a mill that used to produce fertiliser by crushing bones brought by canal.

A house at Hayton has a wall incorporating a row of cartwheels. The church dates from 1120 and has box pews. The Boat Inn also has a children's playground and there is a canalside picnic site.

The most prominent building in Clayworth is the Grade II Hall, a large white mansion that looks down on the canal. The bridge to the north-west of the village is on the line of the Roman road from Lincoln to Tadcaster.

The greenery opposite hides Wiseton Hall. Also hidden beyond Wiseton Park is Mattersey Priory, founded by St Gilbert

Below the A57 at Rhodesia.

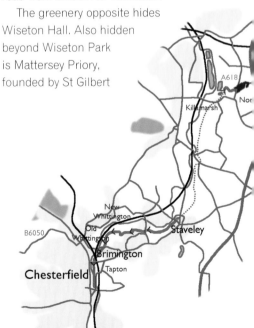

of Sempringham in 1185 but destroyed by fire in 1279. It housed the only wholly English monastic order of the Middle Ages. The Grade II Lady's Bridge or Old Man's Bridge has a bearded face on the keystone on each side.

The canal turns sharply in front of the White Swan Inn with its small basin and wharf. Ahead is the end of the ridge of Cuckoo Hill, carrying the A631, a sandstone ridge that the canal has taken a large detour to avoid as far as possible. The last part of the ridge is passed in the 141m Drakeholes Tunnel, which has no towpath. Travelling north, there is a clear view of any traffic approaching.

Gringley Top Wharf and Top Lock lie below Gringley on the Hill, with its windmill and church. The Beacon Hill viewpoint has views as far as Lincoln Cathedral.

Fountain Hill suggests that artesian conditions have been present as Gringley Carr becomes Misterton Carr, an area drained from marshes by Vermuyden in the 17th century. The spire on All Saints church dates from the 1840s although the church is 13th century and built on the site of an earlier one. The list of vicars dates

from 1254. The stained-glass window in the north wall is over 500 years old.

The cut passes Stockwith Bridge and the Waterfront Inn to enter West Stockwith Basin, where seagoing ships used to offload on to narrowboats. There was also boatbuilding in the basin. Interesting buildings here include the 1797 warehouse.

Trent Lock only opens near high tide and leads out to the River Trent. Just downstream, the River Idle also enters the Trent.

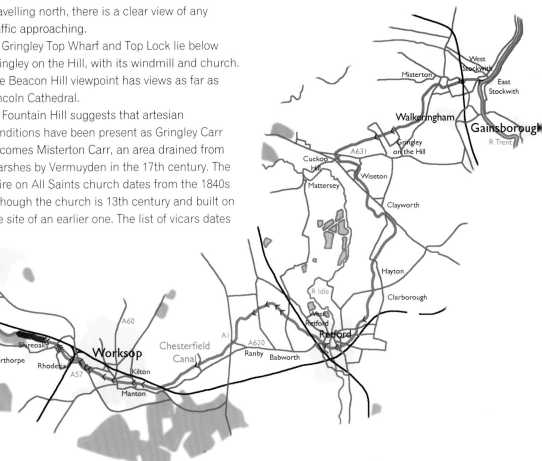

46 Fossdyke Navigation

When Britain's canals vie with each other with claims of longevity, they use as their reference point the Bridgewater Canal, opened in 1761. There is, however, one canal with which none of the others can begin to compare. The Fossdyke Navigation was dug by the Romans in about 120 and is by far the oldest canal that is still navigable by larger craft.

The Fossdyke (a tautological name as fosse and dyke both mean ditch) linked the Trisantona Fluvis, the tidal River Trent, with the River Witham, providing an inland route between the Humber and the Wash. While part of its function was for drainage, the fact that, with the Car Dyke and Cnut's Dyke, it linked right through to Waterbeach in a continuous channel, could only imply its use for transport.

The navigation leaves the River Witham at Brayford Pool, overlooked by Lincoln on its limestone ridge, a striking island in the flat Lincolnshire countryside. The city was the Celtic Lindon, the hill fort by the pool. Later it became Lindum Colonia, a Roman walled garrison housing the 9th and then the 2nd Legion in the 1st century AD, before becoming a settlement for retired Roman soldiers.

The Danes used the navigation when invading England. When Lincoln became a Norman stronghold, the navigation was used for importing the stone for building the cathedral of St Mary, begun in 1072. The cathedral is Lincoln's crown, topping the ridge and being visible from

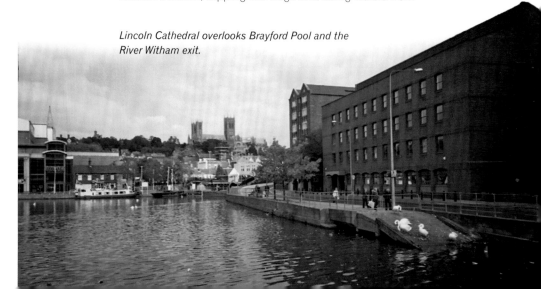

Lincoln Cathedral overlooks Brayford Pool and the River Witham exit.

Brayford Pool and from a considerable area of the surrounding countryside. It was subject to a fire and an earth tremor in the 12th century and has had two restorations. Between 1192 and 1250 it was rebuilt in Early English style, using Purbeck marble. The magnificent central tower was completed in 1311 and has profuse monuments and wood carvings. The Seamen's Chapel has a window depicting Matthew Flinders and the library has a collection of rare books and maps relating to exploration and an original copy of the *Magna Carta*. It was used in filming *The Da Vinci Code*. The Norman castle was built in 1068.

Usage taxed the canal and in about 1121 it had to be scoured out. By the 17th century it was almost impassable again and Acts were established in 1753 and 1762 to improve the navigation. The upgrading in the 18th and 19th centuries of the fenland drainage gave the navigation its present form and it was used commercially until grain barges called for the last time in 1960, since when it has only been used by pleasure craft.

A slipway is located at the east end of the pool in front of the General Accident insurance company buildings. There is a disused lifting railway bridge at the south-east corner of the

Distance
18km from the River Witham to the River Trent
Highlights
The Glory Hole, taking the River Witham
Natural lake of Brayford Pool, overlooked by Lincoln Cathedral
Torksey Castle
Navigation Authority
Canal & River Trust
OS 1:50,000 Sheet
121 Lincoln & Newark-on-Trent

pool. Its neighbouring railway bridge is still very active and the railway line follows the navigation bank to Saxilby, with the A57 also following to Drinsey Nook. The River Witham leaves in the north-east corner under the Glory Hole, Britain's oldest bridge-carrying buildings.

A variety of buildings surrounds the natural lake of Brayford Pool, ranging from modern offices to old warehouses and Lincoln University. The area around the Royal William IV pub has been made into a pleasant waterside pedestrian walkway, continuing along to the public toilets by the harbourmaster's office.

Saxilby faces the navigation.

The Fossdyke heads in a westerly direction towards the Trent. At first there is a line of houses on the right but these are soon left behind and there are very few other buildings alongside the rest of the navigation. Those that there are include traditional brown-brick houses with rib-tiled roofs. On the other hand, the sheetpiling and the raised banks are to be almost continuous features. The former horse racecourse, for which there are revival proposals opposed by residents objecting to traffic, and a golf course, both on the right, are among the first features they hide.

A large watercourse emerges under a bridge from alongside the golf course and crosses the navigation at water level, being lifted out of the other side by a pumping station. Eventually, it finds its way through other channels back into the River Witham and the navigation.

The Pyewipe Inn, named after the peewit, is easier to find by boat than by road, despite being next to the A46 bypass bridge. Other

bridges carry pipes across but the navigation is now in open country for some distance. Occasionally there are willow trees along the banks but more often there is nothing taller than reeds or grass. In addition to passing trains and boats, there are training jets climbing away from the airfield at Scampton.

Burton Waters Marina joins new houses, the Woodcocks pub and a sports centre. The River Till joins under the A57 bridge at Odder.

Saxilby is the only significant village along the Fossdyke and presents an attractive face with mown grass banks. Amenities include the Sun Inn, Ship and a chip shop. It is a village of bridges, being crossed by the A57, a water pipe bridge, the railway turning off for Gainsborough and a neat footbridge brought from the railway in Newark. The arched pipe bridge has handrails leading up one side to the air-release valve at the top, giving an undeserved appearance of dilapidation. Almost below it are the remains of a swing bridge, protected by heavy piling at each side of the navigation.

Torksey Lock with its unusual capstan-operated gates. Tidal water lies beyond.

The windmill on the south side of the village is not seen from the water, just the arms of JCBs in a depot next to the navigation.

The Pike House Inn stands on the left of the waterway. Soon after, a signpost to Doddington Hall, built in 1595 and never sold since, tops the embankment as a tunnel appears under it. The sign is for the benefit of users of the B1190 as the Wigsley Drain diverges from it after 600m and moves off in a more southerly direction.

The navigation turns through a right angle at Drinsey Nook, into a 4km length that is almost straight, more what the Romans might have been expected to build. As it makes the turn, the cooling towers of the former power station at High Marnham are in sight, a scene that is to be repeated. The largest building on this length is a battery chicken farm. Buildings of any sort are spaced well apart.

The banks have deer escape ramps and masses of cow parsley, together with deadly nightshade, blackberries, toadflax, yellow iris and mustard from time to time. The navigation is free of vegetation except for floating weed at the edges.

The navigation's only lock, Torksey Lock, is 400m from the end and acts as the tidal limit, being operable only near high water because of a shallow cill. Its gates are operated by unusual elegant iron capstans. It is possible to carry a small boat right and across the A156, getting in again at a pontoon on the far side of the road. relaunching as before on a set of pontoons where a kingfisher might be found perching. The Wheel House Restaurant is on the left, Torksey itself lies 800m to the north, its stone church tower overshadowed by the brick and stone remains of its 16th century Tudor castle after it was burned by the Royalists in the Civil War. A set of power station towers rise downstream at Sturton le Steeple. These modern edifices aside, it is easy to imagine that the current sweeping along the Trent's wide course has changed little since the days when Torksey was a thriving Roman port linked by a navigation across the flat country to Lindum Colonia.

47 Union Canal

The Forth & Clyde Canal was constructed to bring coal to Edinburgh from coalfields to the west and to allow craft to operate from Glasgow to the Forth and thence to Edinburgh. Its problem was that boats using the Firth of Forth had to be seaworthy and narrowboats were not. Thus, goods had to be transhipped from narrowboats to sailing boats. To get round the problem, the Edinburgh & Glasgow Union Canal was constructed in 1818–1822, linking the Forth & Clyde Canal directly with Edinburgh so that narrowboats could be used for the whole journey. With assistance from Telford, Baird engineered the whole canal east of Falkirk to be on one level. Swift boats were introduced in 1836 to fight back against railway competition, cutting the time to Glasgow to seven hours. The locks were filled in 1933, severing the route. The canal was closed to through navigation in 1965.

Today the Union Canal is complete once again. It runs as a top pound at a single 73m level for 50km from Edinburgh to Falkirk and locks take it down to the Falkirk Wheel and the descent to the Forth & Clyde Canal. As a contour canal, it was nicknamed the Mathematical River.

The Falkirk Wheel in the process of rotating.

welcome to

It was from Gray's Mill that Prince Charlie successfully ordered Edinburgh to surrender in 1745. After the Water of Leith visitor centre, it is followed by the eight-arched Slateford Aqueduct just round the corner, 183m long and 20m high over the Water of Leith. It uses 15m arches, which are the same size as those on the other river crossings, as well as the same iron troughs, which are wider at the invert than at the top. Telford's hollow piers are used although he did not agree with Baird's need for masonry supports to the iron troughs.

A 4m elm sculpture by Robert Coia shows Burke and Hare, two Ulstermen working on the canal. They developed a sideline of selling bodies for medical research but, instead of bodysnatching, they murdered their victims, about 18 in all. Burke was hanged but Hare was released after turning king's evidence.

In 1834, John Scott Russell, the engineer, mathematician and natural philosopher, was watching a barge being drawn by two horses in the vicinity of a bridge at Long Hermiston. The horses and barge stopped but a 300–400mm high wave broke free from the front of the barge and carried on at some 14km/h. Russell followed it on horseback for 2–3km before losing it round a series of bends. As a result of this incident he published his Theory of the Solitary Wave or the soliton, a principle that is now known to have widespread application in the natural sciences.

At Ratho, the Bridge Inn has its own restaurant barge, the *Pride of the Union*. For years this inn was managed by Ronnie Rusack, a prime mover in the restoration of the canal and a promoter of the Seagull Trust.

A 5km lade, the canal's main feeder, complete with tunnels, arrives next to the five segmental arches of stone faced with rock of the 128m long Almond Aqueduct. This carries the canal 23m above the River Almond, on the bank of which is a dinosaur footprint. From here the views are excellent, the Forth road and rail bridges being

Distance
56km from Edinburgh to the Forth & Clyde Canal

Highlights
Edinburgh Quay and Leamington Lift Bridge
Slateford Aqueduct Water of Leith Centre
National Rock Climbing Centre
Almond Aqueduct and dinosaur footprint
Linlithgow Palace and canal centre
Falkirk Tunnel under Glen Village
Falkirk Wheel

Navigation Authority
Scottish Canals

Canal Society
Bridge 19-40 Canal Society
www.bridge19-40.org.uk

OS 1:50,000 Sheets
65 Falkirk & Linlithgow
66 Edinburgh

The Avon Aqueduct, the second-longest of its kind.

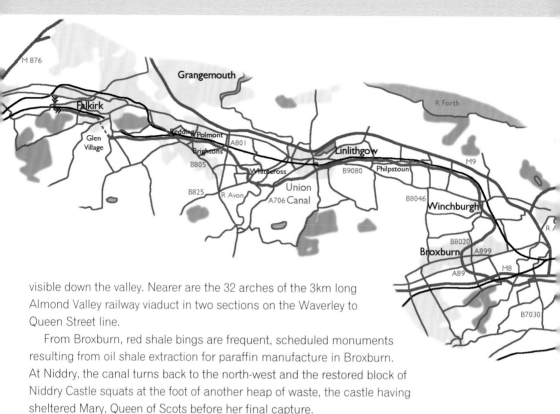

visible down the valley. Nearer are the 32 arches of the 3km long
Almond Valley railway viaduct in two sections on the Waverley to
Queen Street line.

From Broxburn, red shale bings are frequent, scheduled monuments
resulting from oil shale extraction for paraffin manufacture in Broxburn.
At Niddry, the canal turns back to the north-west and the restored block of
Niddry Castle squats at the foot of another heap of waste, the castle having
sheltered Mary, Queen of Scots before her final capture.

The Linlithgow canal basin is a delightful spot in a setting of low canal
buildings with former stables, multi-level bridges and Linlithgow Canal Centre
museum with canal and Roman remains. It is also the home of the *Victoria*, a
diesel-driven replica of a steam packet.

Beyond is the red sandstone block of Linlithgow Palace, built in the 15th
century for James I, birthplace of Mary, Queen of Scots. Cumberland's
soldiers gutted it when occupying it in 1746. The 1960 aluminium crown on
St Michael's church to the right looks strangely out of place, a modern metal
sculpture on a Gothic 13th century sandstone building.

Near the 69ha Muiravonside Country Park is Causewayend, the site of the
Almond Ironworks, bringing passengers from Glasgow and coal from the
Slamannan coalfield to be taken to Edinburgh.

The bridge on the stretch between Whitecross and Brightons is of Armco
culvert, with rubbing strips inside for cruisers and narrowboats. Adjacent to
this bridge is an open area of bank laid out with picnic tables.

Bridges are of sandstone, flat arches meeting vertical abutments almost
at right angles. Each bridge has a projecting stringline just above the arch.
It may be an optical illusion caused by the arch but each of these projecting
courses appears to dip, making the whole bridge appear to sag.

The Glen High Laughin' & Greetin' Bridge is named after the faces on
the keystones, a happy contractor facing the easy run to Edinburgh and the

the next corner on the left bank is the Battle of Falkirk monument. Here Bonnie Prince Charlie beat the English, who were facing driving rain and not expecting the Highlanders to make the first move.

On the right, across Falkirk, is the best view over the Firth of Forth to the hills on the Fife side. A 500m extension added in 1823 shortened the distance passengers had to walk between the two canals.

The 11-lock flight down to the Forth & Clyde Canal was lost under a road. After the Falkirk Wheel Top Locks the canal turns a corner and passes in the recent 168m Roughcastle Tunnel under the Falkirk High to Queen Street railway and the Antonine Wall, known locally as Graham's Dyke. Rough Castle was a Roman fort and a Roman centurion ghost was seen by men constructing the Falkirk Wheel link.

In place of the old lock flight is the iconic 35m Falkirk Wheel of 2002, the world's first rotating boat lift. With 35m arms, it is approached over an aqueduct and lowers boats in its gondolas to New Port Downie. The total load of 1,800t takes just four minutes for the journey, driven by up to four of ten 7.5kW motors. A visitor centre describes this new landmark structure and offers facilities including an exhibition centre, cafe and bar at what is now Scotland's most visited attraction.

Golden Jubilee Lock takes the canal down to join the Forth & Clyde Canal.

miserable face of the one needing to construct the 620m long Falkirk Tunnel under Glen Village. This was necessary as William Forbes of Callendar House would not let the canal pass through his grounds. The only rock tunnel in Scotland, it is wide and takes the towpath through. It is largely unlined so that water drips in places, pouring through in a jet just inside the west portal. Passage through is easy because the far end is clearly visible from the start. Lighting is placed through the tunnel, allowing the limestone features to be seen. Far from taking thousands of years to form, there are curtains, 120mm straws and even longer stalactites that have developed since the tunnel was cut. Sloping rock beds on each side of the exit increase its grandeur. The ghost has been seen of a Glasgow man murdered in the tunnel because of gambling debts.

Lady Kilmarnock had been the unwilling hostess of the Government army's General Hawley in 1746 but ensured that he had plenty of wine ahead of the Battle of Falkirk. Around

The current residents of Edinburgh Quay.

Forth & Clyde Canal

Plans for a canal (originally known as the Great Canal by Smeaton, hoping to upstage the Languedoc Canal) to link the Forth and Clyde estuaries were considered as early as the reign of Charles II. It was eventually to be built from Grangemouth to Kirkintilloch in 1768–1773 and extended to Bowling and Port Dundas in 1786–1790. It was the world's first sea-to-sea route, allowing ships to avoid the north coast. The first major transport project in Scotland, it was finished with money confiscated from Jacobite estates after the failed 1745 rebellion. The line followed a possible former drainage path from the Holy Loch to the River Forth. It was designed for high-mast ships, no fixed bridges being used. The navvies were local men to ease unemployment. Many Roman artefacts were dug up.

The canal carried the *Vulcan*, Scotland's first iron boat, which was used for passenger transport from 1818. It was the birthplace of the Clyde Puffers and also had the first vehicle ferries, carrying loaded carts and railway wagons. A connection to Stirling was proposed in 1835 but never built. It carried coal and iron, passengers in express passenger boats and a series of *Queen* pleasure boats, the last of which ceased operations in 1939. Passenger traffic declined after the Glasgow to Edinburgh railway opened in 1842. The canal was bought by the Caledonian Railway in 1868. The canal carried freight until 1914 and the last vessel passed through in 1962. Closure came in 1963.

The bottom of the Maryhill Flight of locks leads on to the splendid Kelvin Aqueduct with its scalloped faces.

Interest was kept alive with regular trans-Scotland powered inflatable races and the Glasgow to Edinburgh canoe marathon. The whole line was restored by 2001 as the Millennium Link with Millennium Commission funding. Seventy per cent of the Scottish population live within an hour's drive of the link.

The old lock direct from the Clyde is now sealed off but Sealock gives access from the tidal harbour at Bowling, where boats are often laid up at the west end. The upper basin acts as moorings for a variety of small craft, including former fishing boats. The old custom house was built to receive the foreign ships that used to arrive when this was the major transport artery across Scotland. There is a disused Caledonian Railway swing bridge across the canal – a listed monument from which the local youth jump. Although the canal follows the bank of the River Clyde for 5km, it is not seen after the railway bridge. At this point, the river is still quite narrow, despite being used by commercial shipping.

This end of the canal includes several bridges of what are probably a unique design. They are double-bascule bridges, each half lifting separately. The operator needs to fit a windlass that operates a gear train leading to a toothed sector ring and a ratchet wheel, mechanisms from the Industrial Revolution. Crews must have cursed the effort needed to raise them.

Among the greenery is the first Roman fort site at the western end of the Antonine Wall. The Antonine Wall was built 61km from Bowling to Kinneil, circa 142, in the reign of Antoninus Pius. Constructed 20 years after Hadrian's Wall, it was less durable, had a 3.5m high turf rampart on a 4.3m wide stone base with a 12m wide x 1.2m deep ditch, a Military Way road and forts every 3.2km. It was intended to keep out northern tribes at the northern extremity of the Roman Empire. It was attacked twice so the Romans abandoned it and withdrew to Hadrian's Wall.

The feature that cannot be ignored is the

Distance
55km from the River Clyde to the River Carron
Highlights
Erskine Bridge and Titan Hammerhead Crane
Kelvin Aqueduct
Antonine Wall and forts
The Helix and the Kelpies
Navigation Authority
Scottish Canals
Canal Society
Forth & Clyde Canal Society
www.forthandclyde.org,uk
OS 1:50,000 Sheets
64 Glasgow
65 Falkirk & Linlithgow

Erskine Bridge of 1967–1971, one of Scotland's major bridges, carrying the A898 300m over the canal and the river with 55m of clearance to the canal. Despite the great height, the bridge was damaged on the underside by an oil rig being moved downriver from UIE's Clydebank yard to the Captain field in 1996. This soaring structure is, quite literally, on a different plane from the old canal buildings, which stand around its ankles. It is an unusually light structure at 690kg/m^2. Ferry Road swing bridge gave approach to the older route across the Clyde.

The A814 has to be crossed at the north-west end of the Clydebank Industrial Estate by the Dalmuir Drop Lock, built during restoration, Britain's only such lock, lowering boats 2.8m under the road and raising them again on the far side.

Above the buildings by the A8014 can be seen a Titan hammerhead crane, 46m long and 49m high, able to lift 150t. It is on the former site of UIE Shipbuilding, now occupied by Clydebank College. The most remarkable reach on the canal follows. The architects have clearly given free reign to their ideas and the setting is completely surreal. Carefully landscaped gardens lead to

The tidal basin at Bowling, adjoining the estuary of the Clyde.

the pedestrian Clydebank Shopping Centre with two footbridges across the canal. McMonagles fish and chip ship (no, that isn't a typographical error) is the world's first sail-through takeaway, an enormous boat-shaped restaurant, totally out of proportion to the size of the canal, which is built on the bed of the cut.

A railway passes under before the Temple Locks, another runs along the left bank and a third formerly passed under the canal. In the centre of this railway triangle is the Lock 27 where the public house's benches are set out on both sides of the towpath so that it is necessary for those on land to pass between the tables of drinkers.

At the foot of the flight of five Maryhill Locks, ending the 48m climb from the River Clyde to the start of the 26km top-level pound, is the Kelvin Aqueduct, which has four 15m span round arches. Robert Whitworth's 120m long stone structure, with its scalloped faces acting as lateral arches, passes 21m high over the River Kelvin and was the largest structure of its kind in Europe when it was built in 1790.

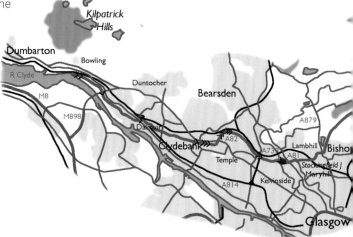

has an unusual number of masons' marks
n the stonework. Contractor William Gibb was
elieved to have made a massive loss meeting
is contract price. A tower block dominates one
ide of the flight. The Kelvin Dock liquor store
verlooks the dry dock. The dock built ships from
789 to 1949, including the first Puffer and D-Day
nding craft. This group of canal features is one
f the finest in any urban setting in Britain.

The Glasgow Branch runs south-east from
tockingfield Junction and formerly connected
ith the Monkland Canal. A floating bridge used
carry the towpath across the junction but this
now routed under the aqueduct. Also gone is
e Maryhill ironworks, its place taken by small
usinesses. From here there are views to the
entre of Glasgow; the British Celtic *glas cau*,
reen hollow, no longer applies.

The canal passes an 1820 church at Cadder,
uilt on the site of a 13th century place of
orship, complete with watchman's house and
ortsafe, an iron coffin to deter body snatchers.

At Glasgow Bridge, a slipway is almost
pposite the Stables public house, former
anal stables, and moorings for
everal boats, including the
anal society's craft.

Adjacent is the high-level concrete A803 Glasgow
Road Bridge on the line of what was once a steel
swing bridge. Construction of this section of
canal plays a central role in Margaret Thomson
Davis' *The Glasgow Belle*. The canal crosses the
line of the Antonine Wall and passes a Roman
fortlet, camp and fort.

From its position to the south of the Campsie
and Kilsyth Fells, Kirkintilloch was an agricultural
and weaving town until the arrival of the Forth &
Clyde Canal, when it became industrialised by
iron foundries as the pig-iron could be exported
by canal. The Barony Chambers were built in
1815 as the town hall. The Auld Kirk of 1644 is
now a museum covering the social and industrial
history of the area, including canal displays.

A recent boathouse is built on a slipway used
when Kirkintilloch was a shipbuilding centre
from the 1860s to the Second World War. It was
another source of Puffers, apparently including
the *Vital Spark*. The 40-berth Southbank Marina

Ancient and modern bridges at Kilpatrick. Even the soaring Erskine Bridge was not safe from being bashed.

is in the former shipyard. The dominant building is the red sandstone church of St Mary of 1912–1914, looking even more impressive because of its position high above the canal. The A8006 viaduct follows.

The aqueduct of 1774 over Luggie Water was an engineering milestone, the first major canal aqueduct in Scotland and the prototype for the Kelvin Aqueduct. Unlike the latter, the navigation channel width is not reduced over the aqueduct. The 38m long x 27m wide structure is 15m high and has horizontal side arches for support. The builders of the Campsie Branch Railway, now a footpath, found there was sufficient room for them to get a line through under the same arch by building a twin-arched culvert for the river and running the railway across this.

Twechar was a mining village. The pivot for a swing railway bridge stands on the right bank of the canal. The railway company was allowed to build its bridge across the canal if it agreed to transport a large proportion of the coal by canal. Hence, this colliery was still moving coal by water when most other collieries had gone over to using trains.

A meandering section of canal at Kilsyth is followed by a swing bridge. Near Kilsyth a soldier on horseback was found preserved in a bog.

Croy Hill has a disused quarry and Roman fort site. Red cliffs surround a picnic site on the opposite side of the canal. Between them is a spur that is Auchinstarry Marina, served by the Boathouse, built by Bristish Waterways' Waterside Pub Partnership at Auchinstarry Marina in 2008 and claimed to be the country's first sustainable public house, including geothermal heating.

The Union Canal formerly connected where some low white railings are located at Port Downie in Camelon. This wharf, above the top lock of the Falkirk Flight, was one of the most important points on the canal, not only as the junction but also because the line is only lightly locked between here and Glasgow. Both canals were served by the Georgian Union Inn. The world's first steam boat was launched on to the canal here in 1789 but had problems with its paddle wheels. The steam boat *Charlotte Dundas* conducted further trials from the Glasgow end in 1802, towing two 70t barges 31km. Experiments were abandoned for fear of the wash damaging the banks. From 1831 to the late 1840s swift boats operated a passenger service that took three and a half hours for the 40km. The boats travelled at 16km/h, pulled by horses that were changed at frequent intervals. Bringing the communications up to date, the towpath between Edinburgh and Glasgow contains fibre optics cables as the first-such British canal communications route.

Falkirk takes its name from the Middle English *faw kirk*, mottled church. Locks are now frequent. The Canal Inn is located below the top lock of the flight. Landscaping, mown lawns and maturing trees make it a pleasant area with a children's playground and bowling greens adding to the welcoming and open feel. A Beefeater restaurant and pub, with canal artefacts, is in a former bonded warehouse. The Grade II Rosebank Distillery, bought by the Isle of Arran Brewery, is in the middle of Falkirk on the opposite corner of the A803, a quite busy road. In the spring there is a spread of purple, gold and white crocuses across a wide area of public lawn. The locks here have iron hooks instead of mooring bollards and the paddle gear is operated by hand spikes.

Schools precede the B902 at Bainsford Bridge in an area that formerly undertook iron casting. In 1298 Sir Brian de Jay was killed here by Wallace's Highlanders after the Battle of Falkirk.

The last 2km of the canal's original line have been filled in and all that remains is a kink in the bank of the River Carron and the main road through Glensburgh, which runs along the former line of the canal. The canal's line was severed by the M9 and A905.

From Ladymill Weir, the Carron Cut runs north as a short branch to join the tidal River Carron beyond a new Sea Lock just above the M9. One benefit of this route is that Skinflats is the best place to see winter wildfowl and waders on the Forth.

This area between Falkirk and Grangemouth is the Helix, with 750,000 trees, paths, cycleways and public art. Displacing the water from the lock are two 35m high, rocking, silver, Kelpie heads, the world's largest equine statues.

Bar Hill at Twechar.

Crinan Canal

The Crinan Canal is isolated from the rest of canal network. It is unlike virtually anything else in Britain. It was built to avoid craft having to face heavy seas off the Mull of Kintyre and cuts 137km off the journey north from the Clyde. In so doing, it provided a market for people in the Western Isles and, in turn, was used to supply them with salt and coal. It formed part of MacBrayne's Royal Route from Glasgow to Oban, named after its use by Queen Victoria in 1847. Even now, it is surprising just how many work vessels on the west coast are still built to fit the canal's lock dimensions.

It was built by a private company launched by the Duke of Argyll. Work began in 1794 under John Rennie but there was inadequate supervision and skilled labour was in short supply in this remote part of Argyll & Bute. Thomas Telford and James Watt became involved in the re-engineering in 1816. It was opened in 1801 with the help of Government loans and completed in 1809. Despite carrying

The terminal basin at Crinan, possibly the most beautiful spot on the British canal system.

heavy traffic, the anticipated profits were never achieved. The year's tally for 1854 was 33,000 passengers, 27,000 sheep and 2,000 cattle. By 1906 it was carrying mainly goods but the situation has switched back with the bulk of the traffic these days being yachts, cruisers and fishing boats. There are no inland craft.

The Maggie, the 1953 Ealing comedy about an old Puffer, was partly filmed on the canal. Clyde Puffers may still be found moored in the basins, Puffers being built to the canal lock dimensions, taking coal to the west coast and bringing back whisky and other produce. An early user was the *Comet*, the first steamcraft to go to sea, wrecked just west of the canal in 1820.

The canal has been called the most beautiful shortcut in Britain. Surely nowhere else on the British canal system can match the sheer breathtaking splendour of the canal basin at Crinan, a basin that has not been ruined by commercialism as might have happened in a more accessible spot. Just a coffee shop and Crinan Hotel, famous for its seafood, have been added to the usual canal basin facilities.

Crinan Harbour lies on the open Loch Crinan. The sea lock at Crinan, as at Ardrishaig, opens at all states of the tide but other locks are only operated from 8.30am to noon and 12.30pm to 4.30pm, Mondays to Saturdays.

Crinan only acquired its name when the canal was opened. Until then it had been Portree, king's port. There are twin sea locks, only one of which is in use, with moorings in the basin for a small number of craft. The bridge keeper's house was named Puddler's Cottage as it was his job to seal leaks in the canal with puddled clay. Lobster pots are stacked up in a garden and a small red-and-white lighthouse acts as a beacon. Beyond all this is a magnificent panorama of islands across the Sound of Jura, Jura, Scarba and Luing.

Above lock 14, a low concrete platform projects into the canal. From here to Crinan Bridge the canal is at its narrowest, cut out

Distance
14km from Loch Crinan to Loch Gilp
Highlights
Canal basin at Crinan
Navigation Authority
Scottish Canals
OS 1:50,000 Sheet
55 Lochgilphead & Loch Awe

of the granite of the hillside. One-way traffic is controlled by telephone. However, between the rock outcrops there are several inlets, lined across with buoys to keep craft from entering but forming useful boltholes in an emergency for small craft.

Crinan Bridge is the end of the one-way section and the start of the B841. This connects by passenger ferry to Crinan Ferry. The canal is closely followed by roads throughout its entire length. Over most of its length it is edged on the south side by the flanks of Knapdale Forest.

The canal has many feeders from lochans in the hills of Knapdale Forest. The one from

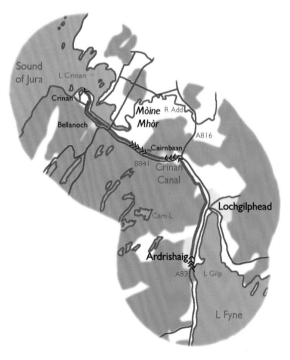

Ardrishaig basin with the Clyde Puffer VIC 27, Auld Reekie, *repainted for the BBC as the* Vital Spark.

Lochan Dùin comes in at low level before the canal begins the climb to its 21m summit level through the five Dunardry Locks. There is a high landing platform before each of the locks, which are separated by large ponds.

The summit pound at 20m is connected to Loch a Bharain. It has the remains of a boathouse that housed the steamer *Linnet*. Almost opposite is the feeder from Daill Loch reservoir, which is flanked by a forest walk. At the far end of the pound there is a rather larger feeder, the sluice-controlled Carndubh Burn, which empties over a weir and under a stone bridge arch, bringing water from Loch na Bric, Loch an Add, Loch na Faoilinn, Cam Loch, Loch Clachaig, Dubh Loch and Loch Gleann a' Bhearraidh reservoirs. Wildlife on this reach, from the mink to the wren, is noticeably tamer than

in other parts of the country. Norwegian beavers were released in 2009, despite concerns that they might spread and damage the canal.

The first four locks down are at Cairnbaan; the fourth, just below the B841 swing bridge, is overlooked by the Cairnbaan Hotel. Near the hotel is Leacan nam Sluagh, the stone of hosting, with Bronze Age cup-and-ring markings. Carn Ban, the white cairn, is on the right after the final lock of the flight. It is a Bronze Age burial mound.

On the far side of the A816, near a fort at Achnabreck, there are more cup-and-ring-marked rocks plus petals, stars and spirals – the most extensive collection of prehistoric rock art in Scotland. Visible from the canal is Stane Alane, a 2.4m standing stone positioned by the wall of a cemetery on the line of the older Lochgilphead coaching road.

Built on the Dippin Burn, Lochgilphead dates back to medieval times but owes its present size

to the canal traffic, being developed from 1790 and the only significant town on the canal.

The A83 is all that separates the canal from the west side of Loch Gilp. There was formerly a wharf at Oakfield. Miller's Bridge of 1877 was named after a bridge keeper in the mid 19th century, who also ran a coal business at the wharf. The manual swing mechanism is located on the west bank so that the owner of Oakfield House could decide when he wanted privacy.

A water-waster of 1895 in a canalside building involves a pair of large buckets on a rocking beam, one of which plugs a drain hole. The device is ingenious but it is not clear how it is better than a simple spill weir.

Above a canalside distillery site was the Robber's Den, on the site of an Iron Age fortress. It was occupied by a MacVicar youth, who raised the alarm when the McAlisters raided the cattle of the MacIvors, precipitating the Battle of Carse.

The McAlisters sought vengeance. When they finally trapped him, they set fire to his hideout and then forced him to jump from the crag on to their waiting spears below.

Ardrishaig was largely created by the canal and still loads local timber, near the ancient chapel of Kilduskland, which was reputedly haunted by its incumbents.

The former Stances Inn took its name from the wooden stances on which fishing nets once dried. Today this sort of activity is long gone. More a seaport than a canal terminus, Ardrishaig has a lighthouse on the 1793 breakwater that established the port. The breakwater was extended in 1932, when a new sea lock was built, the old one being visible beside the opening bridge carrying the A83. There are views across Loch Gilp and Loch Fyne and on a clear day it is possible to see as far as Arran, a canal seascape outclassed only by the one at Crinan.

50 Caledonian Canal

The Caledonian Canal, running north-east right across the Highland Region from the head of Loch Linnhe to the Moray Firth, is unusual for several reasons. Our most northerly operational canal, it is isolated from the rest of the canal system, has large dimensions, incorporates three major lochs and runs through dramatic scenery. It uses the line of Glen Albyn (Glen Mor or the Great Glen), a tear fault line dating from the Caledonian orogenesis, partially infilled with Old Red Sandstone, unlike the thrust faults found elsewhere. Northern Scotland has moved 105km northeast along a wrench fault that has then been scoured by glacier. Without this shift Loch Eil would discharge by Colonsay and Fort William would be near Fortrose. These mountains were once as high as the Himalayas.

The route was considered by Captain Burt in 1726 and it was surveyed by James Watt in 1773. Designed by William Jessop, it was expected to take seven years to build. The three lochs account for two-thirds of its length. This was one of the greatest projects of Thomas Telford, who began the construction in 1803 and

The railway crossing with Ben Nevis in the background

did not finish it until 1822, deepening from 4.6m to 5.2m then taking place before 1847. Capital was provided by the Treasury, Britain's first state-funded transport project. It had been planned to serve two political purposes: it was intended to get potential emigrants used to labouring instead of crofting at the time of the Highland Clearances (but was not entirely successful because of absenteeism at harvest time or when the fishing was good); and it was also to give a safe route for naval vessels away from the risk of attack by French privateers during the Napoleonic Wars, saving 560km on the sea route around the north of Scotland. The risk of attack had declined by the time it was finished.

It reached its most successful point in the 1880s, carrying fishing vessels, grain, salt and timber and running a regular passenger service. Gradually ships outgrew the locks and steamships became more able to round the north of Scotland so use declined. Much use was made of it during the First World War. The main users today are pleasure boats, yachts, fishing vessels and small coasters, with a high proportion of foreign craft. It loses money but is important to the Highland economy as a whole, especially to fishing boats. A 10km/h speed limit operates throughout.

Corpach Sea Loch is the first of 29 locks on the canal, operated during normal working hours from Monday to Saturday, this one only being used within four hours of high water. When it was built, these were the world's largest locks, mostly 55m x 12m x 6.1m deep, and the sea locks are fitted with flotation tanks to assist operation. Entry is made difficult for larger vessels by a flow up to 5km/h across the entrance of the loch on spring tides.

The locks are built as staircases wherever possible and include two of the three longest staircases in Britain. They have horizontal capstan wheels with four ports to take posts for use as windlasses. Drum windlasses are also

Distance
96km from Loch Linnhe to the Beauly Firth
Highlights
Banavie Locks, the longest and widest staircase
Ben Nevis, Britain's highest mountain
Well of the Seven Hands
Urquhart Castle
Loch Ness monster centres
Navigation Authority
Scottish Canals
OS 1:50,000 Sheets
26 Inverness & Loch Ness
34 Fort Augustus
(35 Kingussie & Monadhliath Mountains)
41 Ben Nevis, Fort William & Glen Coe

present by some locks, used until mechanisation in the 1960s. Sea Lock has the first of the lighthouses, small white cylindrical structures with black conical roofs. Indeed, the buildings on the canal are generally black and white, standing out smartly in the rugged scenery.

The staircase of two locks beyond the tidal basin takes the canal up to moorings and its first reach, one of the most dramatic in Britain. Directly ahead is Ben Nevis, the highest peak in the British Isles at 1,344m, 9km away with its summit just 7km from the sea although that summit is frequently shrouded in cloud. Fort William has one of the highest levels of rainfall in Britain, with an average of 239 days of precipitation spread right through the year. An annual race from the town to the summit and back has a record time under an hour and a half.

The first two of the canal's 11 swing bridges are met at Caol. A manually swung railway bridge carries the West Highland line from Fort William to Mallaig with steam trains and observation cars in operation in the summer; the railway line and bridge are now also associated with Harry Potter's Hogwarts Express. It is followed immediately by the A830, the current Road to the Isles.

Banavie Locks are the longest and widest staircase in Britain with eight locks taking the canal up 20m; walls collapsed here in 1829 and 1839. A rebuild was also required in 1929 when a vessel crashed through the top lock gates and fell into the lock below, the sudden water loading damaging the chamber walls. Indeed, the walls were built as the longest pieces of masonry on any canal, each 460m. The flight passes the Moorings Hotel, a craft shop and a quarry. From the top of Neptune's Staircase, as the flight is also known, the view back down Loch Linnhe is stupendous. Among those able to appreciate it are lock keepers whose houses have views down the flight, and those waiting to descend, as boats cannot pass on the flight.

For the next 10km the Western Reach is mostly sheltered by trees, to the extent that the B8004 on one side and the River Lochy on the other are not seen at all. The Upper Banavie aqueduct collapsed in 1843 and Shangan and Loy aqueducts were rebuilt in consequence. Three sluices at Strone act as overflow weirs 1.2m wide and 3m high, the water falling 2.7m to the River Lochy. The turbulence of this resulted in Telford's becoming quite lyrical. Just before the

Quite large ships use the canal. This Swedish vessel is approaching Corpach top lock.

canal passes over the River Loy, there are forest walks in a fragment of the old Caledonian pine forest. The crossings of rivers are usually only marked by groups of stone arches on the left bank or sluices on the right although aqueducts usually have side arches for farm access, not obvious from above. This one has three parallel 76m tunnels, the centre one 7.6m wide and the outer ones 3m wide, all of the order of 4m high. One is for access.

From here, the canal opens out into Loch Lochy, 15km long and 160m deep, the very essence of Glen Albyn as it cuts dead straight between shores rising steeply for hundreds of metres. A reputed 9–12m monster, Lizzie, seems even less likely than the one further north. Canal construction included raising the top water level of the loch by 3.7m. As the canal enters the loch in the natural course of the river, the River Lochy leaves in a new Mucomir Cut to pick up the River Spean at Bridge of Mucomir. A hydroelectric power station was added at the confluence with the Spean in the 1960s.

Beyond the Laggan swing bridge, carrying the A82, the canal enters Loch Oich, the most attractive of the lochs used by the canal. The highest point on the canal at 32m and 6km long with several islands, it is only 50m deep at its deepest point, generally much less, so Jessops designed the earliest type of continuous bucket steam dredger to deepen the channel.

A picnic area (and a grocery shop) surround Tobar nan Ceann, the Well of Seven Heads. The well is topped by a pyramidal monument inscribed in English, French, Gaelic and Latin, the former towards the loch, surmounted by a hand holding seven heads. It was erected in 1812 by Macdonnel of Glengarry to recall the washing of the heads of seven members of the family of the 11th Chief Macdonnel, killed in the 1660s as a reprisal for their undertaking of the Keppoch Murders of their two other brothers.

Inchnacardoch Forest

Invergarry Castle was once the home of the Macdonnel Chiefs of Glengarry but it was burned by the Duke of Cumberland in 1746 after Culloden. Colonel Alexander Ranaldson MacDonnell opposed construction of the canal because he feared it would compromise his privacy, subsequently dying of injuries sustained when the *Stirling Castle* grounded in Loch Linnhe in 1828, six years after the canal was completed. Now the L-shaped tower ruins gaze over the Glengarry Castle Hotel, itself having interesting architecture.

Unlike most other British canals, the Caledonian Canal has limitless water supplies. At Invergarry it is the River Garry that enters from Loch Garry past fragments of the old Caledonian pine forest. Beyond a boathouse, the Calder Burn enters at Aberchalder, opposite a rocky outcrop with heather and pines.

The village of Kilcumin was established by St Chumein, a follower of St Columba, who set up a church. After the 1715 rebellion,

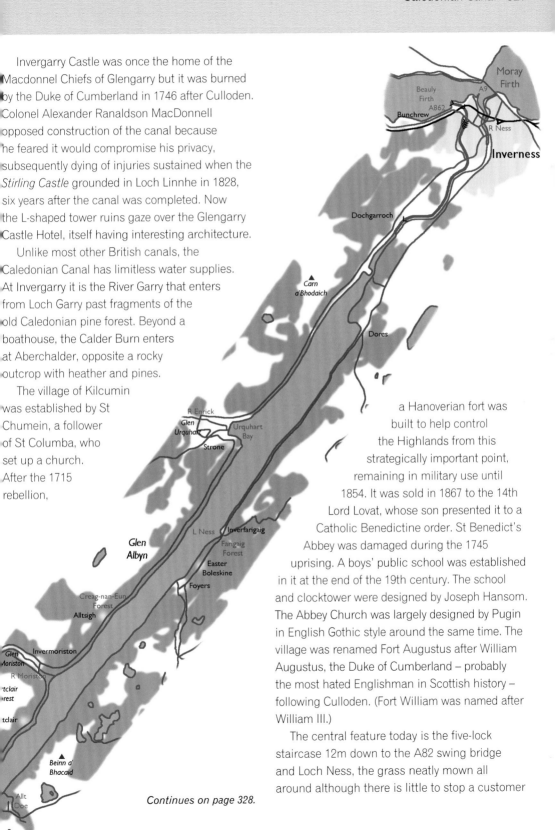

a Hanoverian fort was built to help control the Highlands from this strategically important point, remaining in military use until 1854. It was sold in 1867 to the 14th Lord Lovat, whose son presented it to a Catholic Benedictine order. St Benedict's Abbey was damaged during the 1745 uprising. A boys' public school was established in it at the end of the 19th century. The school and clocktower were designed by Joseph Hansom. The Abbey Church was largely designed by Pugin in English Gothic style around the same time. The village was renamed Fort Augustus after William Augustus, the Duke of Cumberland – probably the most hated Englishman in Scottish history – following Culloden. (Fort William was named after William III.)

The central feature today is the five-lock staircase 12m down to the A82 swing bridge and Loch Ness, the grass neatly mown all around although there is little to stop a customer

Continues on page 328.

staggering out of the Lock Inn in the dark after a dram too many and falling into one of the large lock chambers. Construction had to be undertaken with the stone at the base laid on moss to prevent sand being blown upwards, the base of the bottom lock being 7.3m below Loch Ness water level.

Landing pontoons are adjacent to the entrance to Loch Ness. Loch Ness is Scotland's most famous loch. It is 35km long, covers 56km² and has a catchment of 1,777km². It contains the greatest volume of freshwater in Britain, is deeper than the North Sea at 250m and has the greatest mean depth of any British lake at 130m. Its length has been swum, singly and doubly. The shores are steep with underwater cliffs and the sea rises quickly with fierce squalls, 1.5m waves not being unusual. Thus, it needs to be treated with great respect in bad weather, particularly when the wind is the prevailing south-westerly that blows straight up the loch. The wave action and steep bed and sides mean that wildlife is not as extensive as on the smaller lochs but the water is clearer and it has never been known to freeze, the temperature remaining fairly constant at 5–7°C, especially at the bottom.

The Allt Doe descends over a waterfall to enter the loch on the right side. A sheet of scree slides down from 555m Beinn a' Bhacaidh, facing across to 607m high Burach. The latter is surrounded by Portclair Forest.

Caravans at Rubha Bàn precede Invermoriston. General Wade's bridge over the River Moriston was built after the 1715 Jacobite uprising. The Seven Men of Moriston sheltered Prince Charlie for a month in 1746. Jacobite Roderick MacKenzie, who looked like the Prince, was shot by Cumberland's men in Glen Moriston, allowing the Prince to escape. Opposite, the bed of Loch Ness reaches its steepest, dropping 199m in 110m.

Looking down Neptune's Staircase towards Fort William and Loch Linnhe.

Conspicuous on the east side is an aerial before Foyers. Burns recorded in verse the Fall of Foyers with its 9m and 27m drops on the River Foyers. In 1895 it was used to drive the first large hydroelectric power station in Britain for the British Aluminium Company, now converted to a 300MW pumped storage scheme, using Loch Mhor, 179m above Loch Ness.

A vitrified fort at Strone Point was used as the site for the Grants' Urquhart Castle. At one time it was one of the largest castles in Scotland, commanding the junction of Glen Urquhart with Glen Albyn. Mostly post-1509, it has additions by John Grant of Freuchie, who received it as a gift from James IV. A royal castle in the days of William the Lion, it was fought over in the Scottish Wars of Independence in the 13th century, Edward I capturing it and then losing it to Robert the Bruce. It was sacked in 1689 and, in 1692, blown up by William of Orange's troops to keep it out of Jacobite hands.

It is hard to think of Loch Ness without also thinking of Nessie, the Loch Ness monster. Traditionally shown as a huge serpent with a series of coils looping out of the water, recent drawings suggest a more spherical body with a long neck and small head, perhaps also with four flippers. Recorded sightings go back to a kelpie seen by St Columba in 565. St Adamnan, monk of St Columba and Bishop of Iona, and the monks

of Fort Augustus Abbey, the successors of St Columba, also recorded sightings.

Until 1935, General Wade's Military Road up the east side of the glen was the route through but the opening of the A82 brought tourists with many more sightings and photographs. The Japanese have even searched the depths of the loch with a mini submarine. As yet, conclusive proof has not been forthcoming. Allocation of the scientific name *Nessiteras rhomboideus* seems premature. Perhaps some of the sightings can be put down to the waves that form momentarily on the loch, which can look very black when seen against the light, as is the case for people on the A82.

The Loch Ness 2000 and Original Loch Ness Monster visitor centres in Drumnadrochit put the case for the monster. The findings of a 20-year scientific study said there are only 20–30t of fish in the loch, so they would not be able to support more than 2–3t of monsters. The surface temperature only rises above 12°C for four months of the year, so the suggestion is that people are probably seeing an occasional cold water fish migrating in from the sea. The preferred option is the sturgeon. It can grow to over 4m long and weigh in at 100kg, possibly getting lost from time to time while trying to find suitable breeding waters.

The west shore rises to 501m at scree-coated Carn a'Bhodaich but the east side of the loch is now lower. The width halves abruptly at Dores with a ridge bearing the Kinchyle of Dores stone

Looking up Loch Lochy on a wild day with large waves despite the relatively small size of the loch.

circle and then Aldourie Castle, nearly at the end of the loch at Bona Ferry. Loch Dochfour leads back into canal dimensions after passing Alban Water.

The River Ness leaves to the right over the long sloping Dochfour Weir and then is followed by the canal for some distance. After Dochgarroch Lock, two sets of powerlines cross. A stone circle and Ness Castle are before and after them, respectively, but hidden in the trees on the far side of the river.

Arrival in Inverness, Gaelic for mouth of the noisy river, is announced by the Inverness Rowing Club on the canal.

Whin Park has the relocated 30m-span Dredge-style General's Well suspension footbridge within it. This now supports a people-carrying model railway. Tomnahurich Swing Bridge, also sometimes known as Bught Bridge carries the A82 across for the last time. An isolated, steep, tree-covered hill on the east bank forms the attractive Tomnahurich cemetery, opposite a golf course. The 17th century prophet Brahan Seer had predicted that fully rigged ships would sail inland this way. Ahead, Craig Phadrig has a forest trail near the site of the 4th century BC double-walled vitrified fort that was the stronghold of the Pict King Brude. The 172m high hill commands wide views of the Moray Firth and Beauly Firth. The heart of Inverness, the Highland Capital, lies only 2km from this ancient fort.

Muirtown Locks form a four-chamber staircase, the steps being hidden on the right in front of the Whitecross Restaurant. The 9.8m drop gives a view across Inverness to the Kessock Bridge, which crosses the mouth of the Beauly Firth directly ahead, carrying the A9.

The lock flight is the central feature of Fort Augustus.

The A862, the former route of the A9, crosses on Turn Bridge. Muirtown Basin follows, a wide basin with commercial wharves. Despite the commercial nature of the basin, the west side consists of neatly mown lawns and forms a pleasant environment.

Clachnaharry Lock has a variety of interesting features, a derrick crane, decorated cast-iron lamp standards and a delicate belltower. Directly beyond it is Clachnaharry swing railway bridge, carrying the main Inverness to Wick line, painted white to reduce temperature effects. It was built to carry the Highland Railway across at 65° to the canal. Mounted on the east bank, its 38m hogbacked wrought-iron girders span 24m over the canal. There is a monument to the 1454 battle between the Mackintoshes and the Munroes.

The gently shelving shoreline would have given access to Clachnaharry only at high tide and the muddy bed of the loch was too soft to support structures. Telford solved the problem by running out twin 400m clay embankments that he surcharged with rocks to displace the mud over the following six months, during which time it sank 3.4m. He then excavated the navigation channel between them, building Sea Lock at the far end.

Urquhart Castle is one of Scotland's most visited castles, both now and historically.

Sea Lock operates within four hours of high water and has a light on a post. This supersedes the little cast-iron lighthouse that formerly served to aid navigation. The flows across the end of the canal are 9km/h on spring tides but there is a slack after the flood from an hour after high water at Dover for an hour. There is another after the ebb from five hours forty minutes after high water at Dover for two hours twenty minutes.

Looking down the Muirtown Flight and across the roofs of Inverness towards the Kessock Bridge and the Moray Firth.

Index

ROYAL BOROUGH OF GREENWICH

Blackheath Library
Old Dover Road, SE3 7BT
020 8858 1131

RE

Please return by the last date shown

(8.15)		
1 0 SEP 2015		
ch- 4.1.17		
28 IAN 2017		
WE-21/6/18		
1 3 JUL 2018		

Thank you! To renew, please contact any
Royal Greenwich library or renew online at
www.better.org.uk/greenwichlibraries